M000294835

THE **PHYSICS** OF
KRAV MAGA

THE **PHYSICS** OF

KRAV MAGA

John Eric Goff

JOHNS HOPKINS UNIVERSITY PRESS

Baltimore

University *of* Lynchburg   This book was brought to publication through the generous assistance of the University of Lynchburg.

© 2019 Johns Hopkins University Press
All rights reserved. Published 2019
Printed in the United States of America on acid-free paper
9 8 7 6 5 4 3 2 1

Johns Hopkins University Press
2715 North Charles Street
Baltimore, Maryland 21218-4363
www.press.jhu.edu

Library of Congress Control Number: 2018964634

A catalog record for this book is available from the British Library.

ISBN 13: 978-1-4214-3161-1 (paperback)
ISBN 10: 1-4214-3161-0 (paperback)
ISBN 13: 978-1-4214-3162-8 (electronic)
ISBN 10: 1-4214-3162-9 (electronic)

Special discounts are available for bulk purchases of this book.
For more information, please contact Special Sales at 410-516-6936
or specialsales@press.jhu.edu.

Johns Hopkins University Press uses environmentally friendly book materials, including recycled text paper that is composed of at least 30 percent post-consumer waste, whenever possible.

Like my life, this book is dedicated to Emily and Abby

Contents

Preface

I wish to offer a few words here about this, my second book. I also want to thank a few people who made this book possible. Like many readers, you want to get through the Preface as quickly as possible and jump into the meat of the book. Good for you. I'll be as brief as I can.

This book combines two great loves of mine: physics and sports. My sport of choice for this book is Krav Maga, specifically the style known as Warrior Krav Maga as developed by Chief Master Bill Clark and Master Erick Alfaro at the *Warrior Krav Maga and Kickboxing Center* in Jacksonville, Florida. I'll introduce you to Krav Maga in Chapter 1. For now, I'll just tell you that I truly enjoy doing martial arts, and trying out Krav Maga for the first time while working on my black belt in karate was one of the best choices I ever made. I'll never be mistaken for the next great martial artist or MMA fighter. I try as hard as I can in class, but I have a career as a physics professor at the University of Lynchburg that keeps me very busy. My favorite job is being a dad to two wonderful daughters, a job that I cherish, despite also keeping me busy. So I don't have loads of time during the week to work out and train. But that means that you and I have something in common. If you're not a scientist, you don't have loads of time during your week to learn a bunch of physics. That's what I'm for and that's what this book is for. I will introduce you to key ideas and concepts in physics that will help you understand why Krav Maga techniques are so effective. As much as I love training in Krav Maga, I am *not* promoting that martial arts system as superior to all others. If you take karate or judo or any other martial art, you can apply what you learn in this book to what you practice. The laws of physics where I train in Krav Maga are the same as where you train in a different martial art.

Let me tell you what this book is *not*. It's not a textbook. Don't expect to find a drudging sequence of physics topics like you find in the overwhelming majority of introductory physics textbooks.[1] I have purposely kept equations out of this book. Occasionally I'll convert one unit of measurement to another if I want to look at it both ways, but you'll not find physics equations in this book. My goal is to connect with a reading audience that may not be acclimated to a bunch of mathematics. And believe me when I tell you that I'm not "writing down to you" by omitting equations. As any physics graduate student will tell you, equations are a wonderful crutch to lean on when you don't fully understand the conceptual side of physics. Writing about physics concepts is much harder than leaning on a slew of equations. If any of you are disappointed that there aren't more technical details, good for you. I like it when people want to learn more. I also like communicating science to the general public. If you have questions, please contact me.[2]

This book is not a Krav Maga training manual. Though I take Krav Maga classes, and we have quasi-structured classes and weekly material to learn, I chose not to write a Krav Maga training manual because that's not my specialty. So if the physics content and Krav Maga content aren't presented in a special order, what does that mean? It means you are going to be thrown into the deep end of the physics pool. If you sign up for a Krav Maga class, your first class likely won't be the page 1 material in the Krav Maga training manual. You may be with beginners, but you'll be jumping into the curriculum where the

1. Feel free to find an intro physics book and read in further depth than what you'll find in this book. Even reading about physics topics on Wikipedia for free will work wonders for you.

2. E-mail me at goff@lynchburg.edu and/or stop by my office at the University of Lynchburg in Lynchburg, Virginia (US) if you happen to be in the area.

rest of the class happens to be at the time you join. This book was written so that you should be able to pick it up, open to any page, and start reading. An attacker won't come at you with an ordered set of moves, just as you practiced in Krav Maga class. You'll be shocked, disoriented, and you'll have to fight back. I won't be quite as harsh, but I'll be throwing physics at you from all angles, and I'll repeat myself—a lot. Repetition is a powerful way to learn. The physics in this book will never be more sophisticated than what's seen in an intro physics course. And don't expect a particular technique in Krav Maga to be described by a single physics principle. The laws of physics give us wonderful descriptions of a complicated world. *All* laws of physics could be used to talk about a particular Krav Maga technique. I could easily move a technique in one chapter to another chapter. But I did try to match techniques with physics ideas that could easily be used to highlight an important aspect of a particular technique.

On the pedantic side, I stick mainly with units that we in the United States use. That means that when I calculate some numbers or show a graph, I'll mostly use things like "pounds" for force, "feet" for distance, and so on. I make my living, as almost all scientists around the world do, using the metric system of units.[3] But the United States is, unfortunately, one of the last countries to hold onto English units.[4] I suspect my reading audience will be primarily in the United States, which is why I chose the units I did, but I won't complain if people outside the US buy the book.

Finally, you may see an icon like this ▶ with some of the figures in the book. This is your cue that a video is available to

---

3. The modern name for the metric system is the International System of Units (SI).

4. Though there are differences, I've no interest in splitting hairs between English units, Imperial units, and US customary units.

further demonstrate the technique. You will find the videos at https://jhupbooks.press.jhu.edu/title/physics-krav-maga.

◪ ◪ ◪

I now turn to the acknowledgments I wish to bestow on the people who helped me during the more than four years it took me to create this book. Deserving of my thanks first and foremost is Clif Eli Abercrombie. My family began karate classes in the spring of 2012, and Mr. Abercrombie proved himself to be a top-notch instructor. When he started evening Krav Maga classes in the early fall of 2013, I couldn't resist giving them a try. I became hooked, and by the spring of 2014, I was sufficiently inspired by what Mr. Abercrombie was teaching in those classes that I knew I had to write a book on Krav Maga physics. Mr. Abercrombie possesses a keen intuitive physics of martial arts, and I have learned a great deal from him. I thank him for his patience, high-quality instruction, and willingness to allow me to film numerous Krav Maga techniques.

While Krav Maga classes were being held at *Super Kicks Karate* in Forest, Virginia, Cody Davis, Jessica Maupin, and Nathaniel Hall assisted with the filming of techniques. I thank them for their help. I also thank Whitney Poole for help in filming during the early stages of this book's formation. After Mr. Abercrombie opened *Warrior Success Academy* across the street from *Super Kicks Karate* early in 2018, instructor Janine Koehnke was helpful during a couple of late-book-stage filming sessions. Thank you, Ms. Koehnke.

I thank my faculty colleagues at the University of Lynchburg for their support, especially Will Roach and John Styrsky, both of whom offered insightful comments during several conversations. Thanks to university photographer John McCormick for help with a couple of photos late in the book's creation. I thank university president Ken Garren and provost Sally Selden, both of whom have been terrific supporters of my sports physics work. I thank physics students Brian Ramsey, Chad Hobson, and Nick Savino for their help with filming techniques.

The people at Johns Hopkins University Press have been wonderful all throughout the process of putting this book together. I thank my first editor, Vince Burke, for helping me get the project in motion, and I thank my current editor, Tiffany Gasbarrini, for her efforts in bringing the book across the finish line. I thank managing editor, Julie McCarthy, for her great work through the production of the book. I thank production editor Hilary Jacqmin, editorial assistant Esther Rodriguez, and Ben Kot of Kot Copyediting & Proofreading for all their hard work. I thank Tracy Baldwin for designing the book, and I also thank Katie Feild, Kyle Kretzer, Jennifer Corr Paulson, and Martha Sewall for all their help during the design and production stages. Thanks also go to Rebecca Rozenberg, Kathryn Marguy, and Kalina Hadzhikova for their efforts in promoting it.

I wish to offer special gratitude to my two copy editors, Jim Reilly and Liz Radojkovic. Thanks, Jim, for helping me polish my book prior to submitting it to my publisher. Thanks, Liz, for your meticulous efforts in helping a humble author feel like he has produced a good book. I continually try to improve my writing. Jim and Liz, you both taught me a lot, and I'm grateful. I am solely responsible for any remaining typos or errors in this book.

Halfway into the four-and-a-half years it took me to create this book, my life was destroyed. As much as I would like to leave the previous sentence out of this book, I must include it because of the context it provides for what follows. I offer heartfelt thanks to former student and current physics colleague, Crystal Moorman. I not only thank Crystal for helping me with the filming of some Krav Maga techniques, I thank her for her friendship and the help she provided me during some of the darkest days of my life. I offer special thanks to my family. Their love and support kept me working on this book during a time when I thought I couldn't finish it. My final—and most important—thank you goes to my daughters, Emily and Abby. This book is dedicated to those two wonderful young ladies. Though they won't fully realize it until they are adults, they carried me through unfathomable betrayal. I'll never be able to thank them enough.

THE **PHYSICS** OF
KRAV MAGA

A Strong Mind in a Strong Body

After deciding to work a few overtime hours, you leave your office building and notice that yours is the only car left in the dark parking lot. Only 30 feet from your car, you hear someone approaching you from behind. A quick glance over your shoulder sends a chill down your spine as you see a man with a knife in his hand. He's almost upon you and shouting for help is futile.

How do you think you'd react in such a situation? Would you freeze? Panic? Run like the wind (if you can)? Would you hold your hands up, resigned to being robbed? What if robbery isn't what the guy has in mind? Do you think you could fight back?

There are no easy answers to such questions. Even those of us who have spent a few years studying martial arts might struggle to remember the training we've had. I have not been in many fights in my life, and I'm lucky enough never to have been in the situation I described above. But I also have not studied martial arts long enough to feel that my reflexes would be automatic in such a harrowing situation. I have more work to do.

What is automatic for me are the mental reflexes I've developed after more than a quarter century of studying physics. My understanding of the laws of physics partially makes up for a not-so-flexible, middle-aged body while I'm learning martial arts. Will an understanding of physics help me if I'm attacked by someone wielding a knife? I don't know, and I hope I'll never have to find out. What I do know is that familiarity with the laws of motion adds to the confidence I feel while training. I hope that you'll finish this book comfortable enough with basic physics that your confidence will be elevated while you train.

Before moving on to the next section, I need to plant in your mind a thought about training. Do you play a sport? If you're a golfer, do you think you could become great at the game by sitting on the couch and watching Justin Thomas play golf on television? If you're into basketball, could your game drastically improve by watching highlights of Steph Curry? Do you play a musical instrument? Would sitting in an audience watching and listening to Yo-Yo Ma play the cello or Yeol Eum Son play the piano or Neal Schon play the electric guitar make you significantly better at playing one of those instruments? You obviously know where I'm going with all this. Getting good at something requires lots of *practice*. If you want to be a professional in a given field, expect to practice for many *years*. You won't walk out of a single Krav Maga class with all the skills needed to defend yourself in any given confrontation. That's why it takes years to earn a black belt in Krav Maga, something I've been working on during the writing of this book. You also won't be a physics expert after a single physics class or after reading this book. Luckily, you and everyone else has what it takes to think like a scientist. When we were very young, we didn't shy away from asking questions, a skill every scientist must possess. As most people age, they resist asking questions because they don't want to look like they don't know anything. Ignore that question-asking reticence. Removing ignorance is how we learn. Ask questions in your Krav Maga classes. Don't be shy. Feel free to drop me an e-mail and ask me a physics question or two. And as you seek to understand and practice, play and have fun. Try out what's examined in this book. Come up with your own examples and look at the world around you for other ways to think about the physics I discuss. I'm not suggesting that you spend hours each day trying to reach the level of a professional physicist. I am, however, suggesting that you play with the physics ideas in this book enough to have a passing understanding of why Krav Maga techniques work.

Warrior Krav Maga and Physics

There are numerous martial arts systems you may explore. Thanks to my two wonderful daughters, I began studying karate at age 41. I've now earned my black belt, however, karate isn't the focus of this book. After more than a year studying karate, I began taking Krav Maga classes. Those classes are a *lot* of fun. In Hebrew, Krav Maga means "contact combat." Unlike the stylistic katas[1] performed in karate, Krav Maga emphasizes fighting chops needed in realistic situations. Violent counterattacks and close-quarters fighting skills form part of a Krav Maga student's repertoire. You must be prepared to get hit in an actual confrontation. The key is to give a whole lot more than you receive.

I'm not going to summarize the historical background of Krav Maga. For those interested in that, try, for example, a book by David Kahn,[2] or even Wikipedia. This book is about the physics behind Krav Maga. Will learning some basic physics make you better at Krav Maga? I can't answer that definitively. But I can tell you that there have been moments in my training when a flash of insight resulted from my knowledge of physics. That's a great feeling. It's likely that the best martial artists are not trained in physics, but I bet that they have an intuitive sense of the physics connected to their nearly flawless techniques.

Like all martial arts systems, Krav Maga relies on training students with repetition so as to develop muscle memory.[3] Just as someone won't be trying to remember a given martial

1. The word "kata" is Japanese for "form." A kata is a structured set of movements that can be practiced alone or as part of a group.

2. Two books I especially like from Kahn are *Krav Maga: An Essential Guide to the Renowned Method—for Fitness and Self-Defense* (St. Martin's Griffin, 2004) and *Krav Maga Defense: How to Defend Yourself Against the 12 Most Common Unarmed Street Attacks* (St. Martin's Griffin, 2016).

3. I know there really isn't such a thing as "muscle memory," but I'm sticking with a colloquial definition of the term nonetheless. The term conveys an idea without having to add several more paragraphs on the brain's biochemistry.

arts lesson when confronted with an assailant, no one is thinking about Newton's Laws while being attacked. Imagine, however, developing an understanding of what makes a given technique so effective. Instead of simply mirroring a technique that
one is taught in class, understand why that technique survived
against competing techniques as Krav Maga evolved. In the
language of evolution, some Krav Maga techniques have been
selected based on their effectiveness. Even if physics was not in
the minds of those who developed Krav Maga, the survival of
the techniques in the evolution of the system hinges on the fact
that everything is constrained by the laws of physics. If a new
technique requires too much force or torque to execute, that
technique will be tossed out.

Don't be intimidated by the fact that Israeli special forces
train in Krav Maga. Warrior Krav Maga combines the key ele-
ments of Krav Maga with kickboxing and other physical fitness
specialties. An amalgam of "whatever works," Warrior Krav
Maga is specially designed for those of us who can't devote a
great deal of time to studying martial arts.[4] We learn the main
techniques in Krav Maga and couple that learning with physical
fitness training. Many fights are lost when one person simply
gets too tired to keep fighting. Warrior Krav Maga gets you
physically prepared to never stop fighting if the unfortunate
need ever arises. Two or three classes a week will do wonders
for your physical fitness level and for your confidence in confronting dangerous circumstances. The version of Krav Maga I
take has the "Warrior" label, but I'll simply refer to "Krav Maga"
throughout the rest of this book.

Don't be intimidated by the physics in this book either.
Though physics is a challenging subject to study, you probably
know a bit more than you think you know. But just like turning

4. What's great about the upper levels of Krav Maga is that "whatever works"
turns into "nothing works." Training in the upper levels assumes that the people you
are fighting are highly trained, so you have to be prepared for your initial attacks to
fail.

a couch potato body into a physically fit person, learning some physics doesn't happen overnight. Read a little of this book, then think about it, participate in a martial arts class, and then think again on some of the things you've read here. Over time, physics concepts will sink in and you will begin to see physics everywhere you look. Remember to practice and play with the physics you learn.

What about Math?

Like it or not, mathematics is the language of the universe. Physics is the poetry. Cause and effect are neatly expressed in mathematical equations. Because time has this pesky tendency to tick away and because nobody is content to stay in any one place, mathematics is needed to accurately describe how objects move. Learning mathematics is similar to learning a foreign language. Math has rules that cannot be broken and, when used properly, enlightens humanity to previously unthinkable levels. You will enrich your mind and see the world in a whole new way if you take the time to learn mathematics. If you are a US citizen, learning mathematics will help the country climb up from the embarrassingly low place it occupies in the world's rankings of mathematical literacy.[5]

Now, given what I just wrote, I will keep mathematics essentially out of this book. That's not because I dislike math. I love it, and if you ever drop by my office and ask me for more details behind what's in this book, I'll make a piece of chalk sing across my blackboard with as much math as your heart desires. My goal is to provide you with a conceptual understanding of the physics behind Krav Maga. I'll be tickled to no end if, after reading this book, you want to see some math. For this initial foray into Krav Maga physics, however, I don't want you turning

---

5. A 2015 Pew Research Center study found that the United States ranks 38th out of 71 countries studied in mathematics. The same study had it 24th on the list for science. Given the resources the country has, those numbers are humiliating.

away from our beautiful understanding of nature simply because an equation appears on a page. The hypothetical knife-wielding attacker I described at this chapter's start is scarier than any equation, but I'll do my best to dissect what's needed on the physics side to stop the attacker without recourse to elegant mathematics. If a lack of math disappoints you, you're probably a lot like me. But my years of teaching and media appearances have taught me something basic. It's much more difficult to successfully convey physics in a conceptual way than in a quantitative way. Believe it or not, there is often comfort and safety in working through mathematics with its rigid rules. Trying to convey the meaning behind physical models using words and phrases that are common among nonscientists is quite challenging. I hope I'm successful in what follows.

Intuitive Physics

I suspect that a few of you math nerd types (like me!) are disappointed with the paragraph that closed the previous section. Much like a Krav Maga student being trained to disrupt the balance of his or her attacker, I want to disrupt the balance in the way a person who loves math thinks. Conceptual understanding of physics may be more challenging than always falling back on mathematical descriptions. I wish to stress the need to appreciate intuitive physics.

There are, indeed, many ways we humans learn things, and books, pencils, computers, and other props found in academia represent just a few of the ways. We all learned about balance and gravity at very early ages. As each of us took our first steps, we had an awareness that missing our second steps meant gravity would "beat" us and we would fall. We had neither the vocabulary nor the analytic reasoning capability to have been able to understand anything about gravity and balance.

There are those capable of performing complex physical movements without the slightest idea what calculus is. Some gifted martial artists hone their craft each and every day, and they've never met a differential equation. Even if Newton's Laws are as foreign to you as the surface of Mars, you probably know something about force, mass, and acceleration. A physicist might need to polish your descriptions and, perhaps, correct some misconceptions, but you are perfectly capable of assimilating key ideas contained in Newton's Laws.

Much of our intuitive physics comes from repeating simple tasks so many times that we never think of doing them when we do them. You know that you have to lift a roll off your plate if you wish to eat it. You also know that taking a small bite and chewing it means you either hold the remaining part of the roll in your hand or set it back on your plate. But you don't do any of that with the word "gravity" at the front of your mind. You don't even think, "Gee, if I let go of this half-eaten roll, it's going to fall down," though you know that's exactly what will happen if you do let go. This example is just one of a countless number I could come up with that illustrates how well you became acclimated to living in the world around you. You function within the confines of the laws of nature, even if you have no clue what those laws are.

What do the words "push," "pull," "force," "energy," "momentum," "range," and "motion" mean to you? We have specific definitions for those words in the scientific community. We get pedantic about such things because as scientists, we need to speak a common language and use terms for which we all agree on the meanings. Working with the same linguistic tools helps science to progress. If someone purports to have created a machine that generates more energy than it consumes, most scientists would immediately discard such a claim because of the well-tested law of conservation of energy. Either the person made a mistake in measuring energy or the person is using a term like "energy" where it's not applicable. We as scientists don't intentionally get pedantic with people about the jargon we use; we simply want to be as clear as possible when discussing science. Glibly using a term like "energy" can lead to all kinds of confusion.

Words from the realm of physics pervade this book, but I'll do my best to offer explanations in the plainest of English. I may use a little math for clarity's sake. Far more important, however, is the idea of cause and effect embodied in so many physics equations. An example is Newton's Second Law, which many know as "force is mass times acceleration." Though true in a sense, those five quoted words require a great deal of explanation.

First, "force" is the *net*, external force acting on an object. That means that *all* forces acting on an object contribute to the net force. Not only does the *size* of a force matter (200 pounds, for example), the *direction* matters (to the right, for example). Adding all the forces together requires mathematics that takes into account both sizes and directions of all the forces. The fact that directions matter means that besides algebra, the mathematics needed to accurately calculate the net, external force on an object requires trigonometry.

The "is" is what you might think of as the equal sign in an equation. It certainly equates two quantities that are mathematically equivalent, but there are also philosophical implications.[6] By that I mean the notions of cause and effect. A net, nonzero force on an object *causes* the object to accelerate, which is the *effect* of having a net, nonzero force on the object. One may also reverse the thinking. If one observes an object accelerating, one may conclude that the object has a net, nonzero force on it, even if it's difficult to determine what force or forces are acting on the object. After all, we can't *see* gravity acting on a kicked ball to pull it back to Earth any more than we can *see* air resistance acting to slow down the ball while in flight above Earth. We do have wonderful physical models for the forces of

6. This is serious! I'm not playing a silly word game like Bill Clinton did in his grand jury testimony and getting you to think about "what the meaning of the word 'is' is."

gravity and air resistance. I need those models when researching how a soccer ball[7] moves through the air.

The word "mass" in Newton's Second Law has a precise meaning in classical physics, and it's *not* "weight." On the one hand, mass is an intrinsic part of an object that does not change, even if the object is shipped off to deep space, far from anything else. An object's weight, on the other hand, disappears when the object is in deep space because weight is the pull on the object by Earth (or, really, any other object, but we're concerned only with the pull of Earth here). Weight, unlike mass, has direction. Earth pulls us *down* toward its center, meaning that one's weight at the North Pole is in the opposite direction as one's weight at the South Pole. For those of you (like me!) who want to lose a few pounds of weight, consider just how weak a force gravity is, despite how much it hurts when we fall on our backsides, or how lethal it is to fall from the top of a tall building. I can easily demonstrate to you how weak gravity is. If you're not standing, please do so. Hold this book open in your hands and continue to read. Everything with mass attracts everything else with mass via gravity.[8] The entire Earth and everything in and on it are pulling on this book. And at the very least, you are breaking even with all that. Accelerate this book upward and you'll *beat* the Earth and everything else, including the Eiffel Tower, the Great Wall of China, and St. James's Gate Brewery, just to name three marvels of human ingenuity. The electromagnetic forces between the atoms in your hands and the atoms in this book are much greater than gravitational forces.

The word "times" is simply multiplication. Whew! An easy one for a change.

---

7. Apologies to those of you outside the United States. During a talk I gave in England several years ago, I held up a soccer ball and referred to it as a "soccer ball." I received some good-natured, and deserved, boos for my unfortunate choice of terms.

8. Though it would be a lot of fun, I'm *definitely* not going to improve upon that statement with a discussion of general relativity.

The word "acceleration" means both size and direction. It is the rate at which velocity, which also includes size and direction, changes with time. Because *any* change in velocity, be it size, direction, or both, implies acceleration, a car going around a circular turn at a fixed *speed* of 50 mph is actually accelerating because the car's *direction* continually changes. Though it's not so easy to see when first encountering these physics ideas, the car's acceleration points from the car toward the center of the circular path.

The way I just discussed Newton's Second Law is the way I'll discuss Krav Maga physics in this book. I want to rely on your intuitive physics as I hone your understanding of the science behind a great martial arts system. Mind and body can indeed work as one. And *both* need to be exercised with lots of practice.

What If I Freeze?

Deep in the middle of our brains is our amygdala. Two chunks of brain matter form the amygdala, with each chunk about the size of an almond. Though there are other functions of the amygdala, our concern here is with the emotional reactions to decisions that must be made quickly. The amygdala is like a danger alarm. Once that alarm sounds, memories recorded during the time of danger are much more robust than memories recorded on mundane days. That's why you have vivid memories of the details surrounding an attack, yet you couldn't remember where you put your keys last week when you thought they were lost. As our amygdala records images, sounds, smells, and so on, time feels like it's slowing down for us. It's as if we've entered *The Matrix* watching bullets fly past us.

Having an alarm in our brains is a wonderful piece of evolutionary help. Our ancestors' chances of survival improved if, upon hearing rustling branches, they quickly entered a state of alarm. Our amygdala releases adrenaline into the blood heading toward our muscles, which helps us prepare for danger. Our blood sugar rises precipitously as our body supplies muscles with energy that can be used quickly. The shakes you get while scared is your body's way of speedily getting that energy-laden blood to various parts of your body. Though several of our reactions to fear cause us even more anxiety, our bodies are really helping us prepare for danger. After all, those who heard sounds in the bushes and either readied themselves for a fight or fled the scene had a better chance of survival than those who passed the sound off as "just the wind . . . and not a hungry saber-toothed tiger."

I just described the "fight or flight" response to danger. Each of us reacts in different ways when our amygdala sounds its alarm. Some will flee the scene, even if there is no real danger behind a spooky sound. Others will brace for a fight. There is actually another response besides fight or flight. One may become frozen with fear. Don't criticize that response as a sign of weakness. A predator's eyes, like our eyes, pick out sudden movements, and can sometimes be distracted by those movements. Ever sit in a restaurant with someone and find yourself annoyed that your dinner companion can't stop looking at a nearby television? Many animals have been saved by freezing and blending in with the background while a predator scans the area.

For those who are new to Krav Maga, martial arts, or any type of fighting, please don't worry if your initial response to a stressful situation is to freeze. Your biochemistry, honed by tens of thousands of years of evolution, is actually providing you with a response that could save your life. There are, of course, situations in which freezing won't help. The attacker in the scene I described at the beginning of this chapter will, at the very least, rob you if you freeze. The point of learning to defend yourself with something like Krav Maga is to keep you from freezing and to provide you with responses if an attacker has other things on his or her mind besides larceny.

And please don't worry if you freeze when I throw a few physics terms and concepts at you. Our powerful brains have developed technologies that allow us to live—mostly—peaceful and comfortable lives. Reaching that level of comfort happened in a short time scale compared to the time needed to develop evolutionary responses to fearful situations. Our amygdalas can thus overreact to situations that are much safer than one in which we are being attacked. We feel stress and anxiety during exams, when we're running late to an appointment, and when our favorite sports team is down in the final minutes of an exciting game. A physics concept on a page in this book can't hurt you, but it's perfectly fine to feel a little anxiety while trying to digest a concept that is foreign to you. Just as we do in our stress drills, you will become "acclimated to the sudden shock of being attacked,"[9] or at least you'll soon get used to thinking about some physics concepts.

Exercise and Food

One of the goals in learning Krav Maga is fitness, both mental and physical. I hope to pump up your mental fitness with what's examined in this book. If you've not already done so, make an effort to improve your physical fitness. Hit the gym a couple of times a week if you can. Take walks, use the stairs instead of elevators and escalators, and get your 10,000 steps in for each day. How far do you walk with those 10,000 steps? Consider that two strides represent a distance walked approximately the same as your height. Suppose you are 6 ft (1.8 m) tall. That means you traverse a distance of about 30,000 ft, or 5.7 miles (9.1 km), with those 10,000 steps. That may seem like a lot for a single day, but a few work breaks during which you walk

9. I will sprinkle a few terms and definitions we use in our Krav Maga classes. We are often asked what is a "stress drill," and we respond with the quoted phrase given above. Learning about stress drills occurs during Krav Level 1 Certification, but we participate in stress drills at all levels. More details may be found in the Warrior Manual from Warrior Publishing, 1400 Millcoe Road, Jacksonville, Florida 32225 (2010). I'll refer to the Manual in other places in this book, but I'll use an abbreviated reference.

around your office building a couple of times can do wonders for getting you to 10,000 steps each day. Stretch before and after you exercise, and that includes walking. Physical confrontations sometimes end when one person is simply too tired to continue. Adrenaline helps us in stressful times when we need to elevate our blood sugar and get our muscles moving quickly. But if we're too out of shape to sustain self defense, we'll become part of the crime statistics.

Our Krav Maga warm-ups are intended to "accelerate your heart rate and increase your breathing."[10] One aerobic exercise I enjoy is jumping rope. It's not too hard on your knees, especially if you jump rope on a padded mat, and it doesn't take long to increase your heart rate while breathing heavily. Figure 1.1 shows me jumping rope.

It usually takes me about two minutes of relatively fast rope jumping to get my heart rate and breathing where I want them to be prior to beginning Krav Maga class. I confess that it took me awhile to get to the point where I can jump rope for two minutes straight. As we age, our knees, back, hips, and so on don't help us as much as they used to. And two solid minutes of jumping rope is a lot harder than it may seem. If you're not an exercise fiend, try jumping rope while staring at a clock or while someone times you. You may just find that two minutes elapses a lot slower than you ever imagined!

A colleague recorded me jumping rope so that I could determine my rope's rotational speed. Simply watching the video and timing a few rotations of the rope would have been good enough for what I need here, but I wanted more. So, I determined my rope's rotational speed as a function of time. After analyzing the jump-rope video frame by frame,[11] I came up with Figure 1.2.

---

10. Warrior Krav Level 1 Certification Manual.

11. I used the free software *Tracker* to do the frame-by-frame analysis. It's easy to use. Those of you with videos of your exercise or athletic achievements can use *Tracker* to go frame by frame and find the perfect image. For more details, visit https://physlets.org/tracker/.

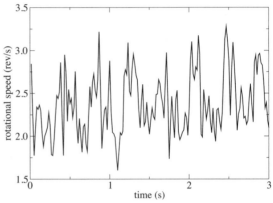

Figure 1.1
Me accelerating my heart rate.

Figure 1.2
My jump rope's rotational speed while jumping rope for 3 s.

The graph in Figure 1.2 shows my rope's rotational speed in revolutions per second (rev/s) versus the 3 s of rope jumping I analyzed. I certainly can't jump rope at a constant rotational speed for my jump rope, but my rope's average rotational speed looks to be about 2.5 rev/s. That corresponds to 150 revolutions per minute (rpm), which is roughly one-third of the rotational speed of helicopter blades. Before you get too impressed with that rotational speed, note that a washing machine spins about twice as fast as helicopter blades. But it's more impressive to compare my jump-rope speed to a helicopter than to a washing machine, right?[12]

Don't forget to eat well. A big part of physical fitness is diet. You don't need me to tell you that cutting back on sweet snacks and desserts is a good thing. You've likely heard that many times, including in elementary school when you first learned

12. It takes a lot more power to get helicopter blades moving than it does to get your washing machine going during its spin cycle. Think about how much more massive helicopter blades are than the well in a washing machine.

about how your body functions. As a physicist, I think in terms of energy units like Calories.[13] It was to our evolutionary advantage to find high-Calorie food because our ancestors may have had to go several days without eating. That is partly why we love sweet foods. Those doughnuts you have with coffee in the morning are packed with Calories. The problem is that when societies advanced to the point where they were capable of providing us with food whenever we wanted to eat, we didn't lose our love of high-Calorie foods. Over a quarter of the US population is obese, according to the Gallup-Healthways Well-Being Index, and consumption of too much sweet food is a big factor in that statistic.

A pound of fat in your body has an energy content of about 3,500 Calories. Think about ways you can eliminate that pound of fat in your body. If I push myself on an elliptical in the gym, I can burn about 800 Calories in an hour.[14] That means I would have to get to the gym four or five days per week and work hard on an elliptical for an hour each visit just to burn a pound of fat in a week. I love Snickers candy bars. A 2-oz (57-g) Snickers bar contains 280 Calories. Suppose I eat a Snickers bar for a snack twice a day, perhaps once in the midmorning when I'm still feeling like I haven't awakened yet, and then again in the midafternoon when I feel like I need some energy to get me to the end of the day. If I were to eat 14 Snickers candy bars in a week, that would represent 3,920 Calories. Which is easier, working hard in the gym five times per week or not eating those candy bars? It was helpful to our ancestors on the African savanna to

13. I capitalize the energy unit because I'm using what's called the "large calorie." That is the food energy unit you see on nutrition labels. Scientists used the "small calorie" unit many years ago in experiments involving heat flow. There are 1,000 "small calories" in a "large calorie." That is why you see kilocalories (kcal) used on nutrition labels in Europe and other parts of the world where the metric system is used.

14. Take energy burn on gym machines with a sizable grain of salt. Some machines simply take energy output and make a conversion to energy burn using an average human burn. Other machines are more sophisticated and require you to at least enter your body weight.

find high-Calorie foods, but it doesn't help us much now. Skip-
ping a candy bar snack is a whole lot easier than burning 280
Calories in the gym.[15] By "easier," I mean that not eating a
candy bar takes far less physical effort than working hard in the
gym. As for mental effort, candy bars are mighty tasty.

How many Calories do you consume in a day? According to
the dietary guidelines at health.gov, adult women should consume 1,600–2,400 Calories per day. For adult men, 2,000–
3,000 Calories should be consumed. The higher end of the
aforementioned ranges should be used by active people,
whereas the lower end of the ranges should be for couch pota-
toes. Getting your 10,000 steps each day will help you burn
energy at the high end of the government's recommended energy intake.

We use the term "power" in physics to mean the rate at
which energy is consumed (or outputted). Suppose you consume 2,400 Calories per day, which is at the high end of the
recommended intake range for women and in the middle of the
range for men. Dividing that energy by one day gives 100 Calo-
ries/hour. Though the power I just gave is indeed in units of
power, those units are not so familiar to you. A power of 100
Calories/hour is the same as roughly 116 watts (W).[16] What
that means is that if you consume about 2,400 Calories per day,
and then burn that much each day so as to maintain your body
weight, you metabolically burn energy at an average rate of a
little more than a 100-W light bulb. Keep your energy intake
close to the recommended range and do your best to get 10,000

15. I am simplifying how Calories are converted to fat and then burned in the
body. Consumed Calories also come in the form of proteins, carbohydrates, and fi-
ber, to name a few. And burning Calories at the gym doesn't simply mean just burning fat cells. But the numbers I gave are reasonable estimates and the conclusion
that it's far easier to lose weight by reducing Calorie intake than burning Calories at
the gym is certainly valid.

16. One must first covert Calories to the metric system's energy unit, which is the
joule (J). The conversion is 1 Calorie $\simeq$ 4.2 kJ, where kJ (kilojoule) is a popular energy unit in Europe, one you'll see on food labels. A watt is defined as one joule per
second.

steps and/or other types of exercise each day. You don't have to be a world-class athlete, but keeping your fitness at a reasonable level ensures that your Krav Maga training will be optimally effective. It will also improve your chances in a potential future assault upon you.

The Greatness of Science

I wish to offer a few words here that praise science, and not just because I am a scientist and I make my living using physics. Think of science as the best tool kit we have for understanding the world around us. Science gives us methods and techniques to achieve that understanding. You may have heard of the "scientific method," though there really isn't such a rigid method as the one you might have been introduced to in school. Discovery sometimes happens during chaotic bursts of imagination and fearlessness in the face of ardent opposition. We scientists should not have agendas when practicing our respective trades. We should always rely on data and evidence to point us in the direction of how the world works. Hypotheses may be made; most will fail, but some will stick around for a while. Appealing ideas are then tested by colleagues, perhaps around the world. Science is powerful in part because of the crucial step of peer review. Sound scientific ideas must be *falsifiable*. That is, an idea must have the capacity to be proven wrong. If I tell you that I have an idea that dropping a ball from a height of 6 ft will take 0.6 s, you know that I've given you a falsifiable idea. You can take my idea and put it to the test. If you have trouble falsifying my idea, I may have stumbled onto something interesting. If I can refine my idea to determine how long something dropped from *any* height will fall before hitting the ground, I could be sneaking up on where Isaac Newton was more than three centuries ago. But if you get some really great measuring equipment, you might find that my 0.6-s estimate isn't quite right. My initial idea may have only used the pull of Earth on the ball, whereas a better idea needs to worry about air resistance. Even better ideas make use of the spinning Earth, Earth's nonspherical shape, other planets, and so on.

Once a really good idea has been published after peer review, challenged and had its results verified by many scientists, and remains without falsification, we scientists may haul out the word "theory." Please don't confuse that word when you hear nonscientific people say, "It's just a theory!" Calling an idea a "theory" in science is like holding its hand up after it pummeled a hundred Krav Maga opponents. There aren't that many theories in science. The "theory of gravity," "germ theory of disease," and "theory of evolution" are among the elite ideas in science. They have survived much scrutiny and testing. We believe them to be accurate descriptions of the natural world because of how well they comport to reality. And if new information comes along to challenge an existing theory, we scientists use that information to determine if that theory needs to be modified or even tossed out. More than likely, new information means refining an existing theory. We're not likely to toss out the "theory of gravity," "germ theory of disease," and "theory of evolution" because they describe reality very well. We may have to refine realms of applicability or make a few tweaks to the ideas that make up a theory.

I will mostly use Newton's laws of motion. They are over three centuries old and describe quite well what we see around us. They are not always intuitive. That's why it took humans tens of thousands of years working on understanding our world before Newton gave us a theory that worked amazingly well. Since Newton's time, scientists have discovered that his laws break down if objects move close to light speed or if there are large gravitating bodies around. Those laws also fail stupendously in the world of atoms. But despite the fact that punches are fast, trainers have massive bodies, and we are all made of atoms, none of the aforementioned realms where Newton's theories break down will be visited in this book. Even though they are more than 300 years old, Newton's Laws will suffice for what we need.

Evolution and Krav Maga

I've already mentioned evolution a few times in this chapter. I wish to put forth a connection between evolution and Krav Maga (or any martial arts system) that surely can't originate with me. Krav Maga is a unique martial arts system, just as every other martial arts system is unique. The laws of physics constrain what all of us can do, which means there is no martial arts system that can employ laws-of-physics-violating techniques. For all the various differences among martial arts systems, they all contain numerous similarities. The reason, of course, is physics. Getting a person to fall over could involve a leg sweep, a clever push, or even a little trickery using a nearby object. Your new understanding of the physics behind Krav Maga will allow you to see why other martial arts systems are effective as well. I guarantee it. There are plenty of ways of punching someone's face, but the person getting hit will know about the concept of force, regardless of which system's punch technique was employed by the puncher.

Physics is the most basic of all the sciences. We physicists study how fundamental particles interact, such as how the proton and electron in a hydrogen atom do their quantum dance. Working with a relatively short list of rules, we can describe with great precision how a lot of the natural world works. Physics is the basis for chemistry, which is the basis for biology.[17] Biology then forms the basis of macroscopic animals and the basis for psychology. And evolution must be understood if anyone wishes to understand modern biology. Heck, any educated person should have at least a passing understanding of evolu-

17. Don't get me started on the absurdity of teaching children science in the order of biology, chemistry, and then physics. I'm definitely a proponent of the *Physics First* educational movement.

tion. We couldn't fight diseases without understanding evolution and, I'm about to argue, we couldn't fully understand why martial arts systems are so effective without a basic understanding of evolution. Don't worry, though, because just as I'm not planning to turn this book into a full-fledged physics textbook, I'm not about to write scores of pages on the topic of evolution. I simply want to toss an idea into your brain before moving on to explanations of the physics behind Krav Maga.

Evolution explains the diversity of life on Earth. The theory of evolution by natural selection provides an understanding of the mechanism by which traits change from one generation to the next. There are various pressures that cause changes to take place. It's thrilling to think about the fact that we share a common ancestor with every living thing on Earth. All animals alive today have evolved to where they are through many generations. No animal is any more evolved than any other animal. Each species has its niche for the environment in which it inhabits. We humans aren't "more evolved" than other animals just because we have big brains and we can take Krav Maga classes. We can't survive on the overwhelming majority of Earth. Just think about how much of Earth's surface is covered by oceans. And just think how fortunate we are to be alive and be thriving contributors to a species that did what something like 99% of all species didn't do—survive.

Now I want you to think of evolution in a different way. We are all members of the species *Homo sapiens*. We really came into our own more than a quarter million years ago when we separated from *Homo erectus*. That might seem like a long time ago, but it's less than 0.01% of the age of Earth. And we're still evolving. Each of us is a transitional form to the next generation. But think about what we are like now. Ever get poked in your eye? It hurts! Ever have your ears popped by someone's hands or another object? It hurts! We humans have plenty of vulnerable areas. There are evolutionary reasons why we have vulnerable areas, some well understood and some not so well understood. Think about all the ways you are strong and protected. Do you have calluses on your hands that could be pressed hard and not cause you pain? Guess what? You won't be learning techniques in Krav Maga class about trying to puncture hard calluses. You *will* learn techniques about striking vulnerable areas. Evolution got our bodies to their current forms. Martial arts systems exploit weaknesses in our bodies. If we had to fight other animals, our martial arts systems would look very different. We don't fare well if we take hard shots to our spines, but it's tough to hit a turtle's spine because of its shell.

As you go through this book, try to imagine how each Krav Maga technique exploits one of our weaknesses. Ask yourself if there was no evolutionary reason for developing a particular strength or protection, why not exploit it? Ever had your fingers pulled apart? It hurts! But there was no evolutionary advantage or need to have strong resistance to having our fingers pulled apart. So, Krav Maga and other martial arts systems come along and say, "Gee, it seems to hurt people when their fingers are pulled apart, so let's add that little bit of harm to our system." I'm not expecting you to work through mounds of equations and perform lots of calculations so that you can understand all the intricate details behind Krav Maga techniques. I'm also not expecting you to become well trained in biology and understand all the anatomical developments we humans have attained via evolutionary processes. But I do think an outside-the-box way of viewing something can be helpful. If you've never thought about how martial arts systems may want to exploit our evolutionary weaknesses, give that idea a little time to stew in your mind. Then read the rest of this book and you might think at some point during your reading, "Oh, I see how that works. And it especially works because we humans don't fare particularly well when we get punched in our throats."

Training Time

Like learning a new technique in Krav Maga, learning a new physics concept can be challenging. You won't execute the new Krav Maga technique perfectly after practicing it just once, and you won't acquire a sound understanding of the physics behind the technique after an initial cursory reading of a few sentences in this book. You may have to try out a technique several times, reread the relevant book section, and then try the technique again before it really sinks in about what's happening on the physics side. But that's no shock, right? Think of anything you do that you do really, really well. You only got good at it by practicing over and over. I keep repeating this point to hammer it home.

We are told in lower-belt classes to "use whatever works."[18] Though there is structure in our classes and a system of belts we test for, we often employ all kinds of techniques we've been exposed to at various levels. I'm going to develop your understanding of Krav Maga physics in the same way. There will occasionally be some structure to the physics I introduce you to, but this is not a physics textbook. I want you to be able to pick this book up, open it to any page, and read about some technique and its associated physics. I'll repeat myself a bit so that I can hammer a few big physics ideas into your brain. Just like Krav Maga isn't a system designed to provide you with some scripted attack or counterattack, this book isn't an academically ordered set of physics topics that appeals to those of us working in colleges and universities. Be prepared for the physics to jump all over the place.

You should now have some idea what this book is about. I hope you're on board with maintaining good physical fitness habits, including eating well. And I hope you're a little less stressed about learning some physics concepts. It's time now to head to the science mat and start training.

---

18. Warrior Krav Level 1 Certification Manual.

Disrupting Balance

Have you ever been talking to someone and that person says something completely strange or shocking? You feel taken aback, and perhaps your wits are slowed enough that you don't have a great comeback. Only later when you're composed and you can think clearly does the perfect response to the person's upsetting remark come to you.

Let me make the situation a little scarier. You are heading toward your neighborhood bar. About 100 feet from the bar's front door you inadvertently bump into a stocky guy who clearly left the bar drunk. The guy thinks you were trying to start something, shoves you pretty hard, and yells, "Let's go!" You weren't expecting this confrontation; you were walking toward a bar with the intention of meeting up with some friends. Seemingly out of the blue, an inebriated and fairly well-built guy has shoved you and wants to fight. Do you have your wits about you? Are you able to quickly switch gears in your mind from being taken aback and shocked to being ready to defend yourself? Do you have the awareness to check out your surroundings, looking for possible help, a path to a place of safety, and, perhaps, friends of the drunk who may be closing in to help their quarrelsome mate?

The situations I described are ones that alter your mental and physical equilibrium. In the case of the drunk, he's attacked you with the intention of putting you in a position of disadvantage.[1] Imagine now that you've trained in Krav Maga. You assess your surroundings, know where to go once you've handled

1. Warrior Krav Level 1 Certification Manual definition: "Self Defense: an attack is initiated towards you and puts you in a position of disadvantage."

the immediate danger, and you can do so immediately after being shoved. Imagine further that the guy comes at you and you react in such a way as to turn the tables on him. You counter fast enough that he's the one taken aback. You've disrupted his equilibrium, caused him to feel taken aback. In other words, you've acted rapidly enough to disrupt your attacker's balance.

What we need to understand is the concept of *balance* from the physics viewpoint. Though you certainly won't be thinking about forces and torques while being attacked, your counterattack may benefit from the basic physical understanding of balance that you developed during your training. Even though I've put the topic of disrupting balance into a single chapter, most Krav Maga techniques, if executed effectively, lead to a disruption of an attacker's balance.

Center of Mass

All matter is made up of atoms. Though we physicists have learned how to deal with the atomic world, and worlds even smaller than that, I have no interest in taking up atomic physics here. Each cell in our bodies has about 100 trillion atoms, and there are about 100 trillion cells in our bodies. That means there are something like, hold your breath, 10,000 trillion trillion (10^{28} if you like scientific notation or 10 octillion if you like showing off your vocabulary) atoms in our bodies. There are, in fact, so many atoms in a typical human that we never even think about the corpuscular nature of ourselves. We're simply continuous "stuff" like skin, hair, bones, blood, muscles, and so on.

All of the stuff that we are comprised of gives us our mass. It turns out that mass and weight are not the same thing, though many people often use the terms and their respective units interchangeably. I'll now elaborate on the mass and weight discussion from Chapter 1. Our mass is intrinsic to us— it won't go away even if we find ourselves unlucky enough to be all alone drifting in outer space. Our weight is the gravitational force exerted on us by Earth. If we go to the moon, we'll weigh about one-sixth as much as we weigh on Earth because the moon's gravitational force on us is about one-sixth what it is on Earth. But our mass while we're on the moon is the same as our mass while we're on Earth. If we ever find ourselves in deep space, we'll still have mass, but no weight because there is nothing around to exert gravitational force on us. Of course, we'll have more pressing things to worry about if we're in deep space. Though "center of gravity" and "center of mass" are technically different, either term will work for our Krav Maga physics.

Think now of being shot from a cannon in the circus. After you leave the cannon, your arms and legs are flopping about in complicated ways. You may even have left the cannon improperly and you're tumbling and twisting. For all that complicated motion, there is one point that is moving along a fairly simple path that even a high school physics student can describe to reasonable accuracy. That point is your center of mass. That imaginary point follows a smooth curve through the air. Its orientation with respect to your body changes as your arms and legs move about, but someone sitting in the bleachers laughing at you, while you worry if the net that's supposed to catch you is tied securely, would see your center of mass tracing out a nice, smooth trajectory (if such a trace were possible).

Center of mass need not even be contained within an object. One of those scrumptious glazed doughnuts that we should avoid like the plague[2] has its center of mass in the middle of the hole. A tossed doughnut might spin about its hole, but the center of the hole would move along a smooth path. High jumpers take advantage of having their center of mass

---

2. Like a lot of people, I'm a sucker for a good glazed doughnut. But thinking about eating fried dough and sugar with at least 225 Calories makes me think about how quickly I can toss away a single Krav Maga class with a few doughnuts.

outside their bodies. The famous Fosbury Flop[3] allows a high jumper to leap head first and bend over the bar, facing the sky, while his or her center of mass passes just under the bar. Given that the law of conservation of energy restricts how high we can elevate our center of mass after jumping, the Fosbury Flop not only revolutionized high jumping, it allowed athletes to sneak up to the very constraints of physics.

Why Do We Fall?

Sometimes your best maneuver is simply getting your attacker to fall. That may be all that's required to run away from danger. Figure 2.1 shows my Krav Maga instructor, Clifton Abercrombie, standing with his back against a wall.

There are three photos in Figure 2.1. The top photo shows Mr. Abercrombie against the wall. The middle photo shows Mr. Abercrombie extending his left leg. The bottom photo shows Mr. Abercrombie falling forward. Try this yourself! You won't be able to stand straight up against a wall, extend one of your legs out, and not fall forward. Even a physically fit and muscular person like my instructor can't keep from falling. By moving his left leg forward, Mr. Abercrombie's center of mass moved from inside his body to just outside his body. I put a star at the approximate location of his center of mass. Balanced against the wall, the star is above Mr. Abercrombie's shoes. As his left leg extends, the star moves in the direction of where his leg mass is moving. Note that the star won't move as far as his leg because only about one-sixth of Mr. Abercrombie's mass is contained in his left leg.

When his center of mass, or the star in Figure 2.1, has moved out past Mr. Abercrombie's shoes, there is a torque provided by his weight that causes him to rotate clockwise in the

3. The technique is named after Dick Fosbury, who won gold in the 1968 Summer Olympics in Mexico City.

Figure 2.1
Watch out for that moving center of mass.

figure. A simple way to think of torque is a force multiplied by a lever-arm distance. The lever-arm distance is from the point of application of the force to the axis of rotation. Only the force component perpendicular to that lever-arm distance will create a torque. A torque is necessary to change an object's rotational state. By moving the center of mass past the shoes, there is now a lever arm for the weight to cause a torque. That is why we like long lever arms on wrenches, and why we put doorknobs far from hinges. I'll have more to say about wrenches in Chapter 5.

Chances are you already know all this physics, even if you don't realize it. A bit more or less than a year after you were born, you took your first steps. You got used to gravity in the months leading up to that magical moment that still brings tears to your mom's eyes when she recalls the memory.[4] You became acclimated to the idea that extending arms and legs outward leads to instability. You then took advantage of that instability and began walking. The simple act of extending your leg forward and letting Earth's gravity exert a torque on your body turned you from a bumbling baby to a highly mobile rug rat.

You can actually get a decent estimate of your own center of mass by doing what Mr. Abercrombie did in Figure 2.1. There are more sophisticated ways to locate a person's center of mass, but it's not crucial that we delve into descriptions of those ways for what we study in this book. When you stand upright, your center of mass is roughly halfway between your front and back at the height of your waist. A standing adult woman's center of mass is about 5/9 of her height, measured from the floor. A standing adult man's center of mass is roughly 2% of his height higher from the floor. Anatomical differences between women and men account for a woman's center of mass being lower in her body than a man's center of mass. The aforementioned com-

---

4. This applies to your dad's eyes, too. My eyes will shed a tear or two when I recall when my daughters took their first steps.

ments and numerical estimates are based on average body types and don't necessarily apply to specific individuals. After all, there are all kinds of body types, from short to tall and gaunt to obese. There are, however, cute demonstrations for showing that a woman is "more stable" because her center of mass is lower than for a man. One of my favorite demonstrations, which I have students do in my classes, is something I first saw as a kid watching a rerun of *All in the Family*.[5] A quick search on YouTube will take you to the video of the Bunkers taking turns lifting a chair with their heads on a wall. The women could do it; the men couldn't do it. It's not a demo that works in the way you think it will every time it's tried because, as I already noted, there are many different body types. But it's fun to try!

Leaving '70s television and returning to Krav Maga physics, even the biggest and strongest attackers can be caught off guard with a disruption of balance that leads to a fall, thus creating the opportunity to flee the danger zone. We are instructed during our Krav Maga classes how to disrupt an attacker's balance and, perhaps more important, how to keep our balance while being attacked.

How to Keep Your Balance

What do you typically do when you trip or something else happens that makes you lose your balance? Do you stick your arms out? Perhaps you even stick your leg out? You may look funny doing all that, but you are redistributing your mass so as to keep your center of mass above one of your feet. If you actually fall, you put your arms out so that they hit the ground instead of your head. Life experience has taught you how to move your arms when you lose your balance. You don't need to do any calculations or think about optimum arm placements. Instinct

5. Season 1, Episode 5, "Judging Books by Covers."

takes over and you do what you can to avoid falling. Krav Maga helps us develop instincts for maintaining our balance and, with lots of practice, we hope to maintain our balance if ever in a dangerous confrontation.

Remember the hypothetical situation of a drunk you accidentally bumped into? What if you bumped into him as he was walking quickly by you, and then you heard footsteps behind you a few seconds later? Perhaps the guy was drunk enough that it took him a second or two to realize what happened and then get mad at you. He then decided to turn around and go after you. What if you turn around and the guy is less than 10 feet from you and closing the distance fast? You have options, but one option is to disrupt his balance quickly, allowing you time to escape. This can be accomplished with a "long-range attack," which Krav Maga defines as "within kicking distance."[6] It won't take long before a guy coming at you from 10 feet away will be within kicking distance.

Think about the mass in your leg. I already told you that about one-sixth of Mr. Abercrombie's mass was being extended when he lifted his leg. Women have slightly more than a sixth of their mass in one leg. You can do the math for yourself, but a 180-lb man has roughly 30 lb of his weight in each leg. Each leg represents the mass of about two bowling balls. Couple that with the rest of the body driving a leg, and getting kicked can really hurt.

So your split-second decision upon turning around and facing your attacker is to kick him. He's drunk and the chances of you landing your kick on your intended target are pretty good. Your intended target should be a vulnerable part of his body, such as a knee or his groin. There is no need to try anything fancy like kicking his head. You don't want to do anything that puts you at greater risk. How are you going to execute your

---

6. Warrior Krav Level 1 Certification Manual.

kick? You know from what we've already discussed that extending your leg out could cause you to rotate forward. That might
actually be something you *want* to happen, if you are executing
an *offensive* front kick. In that case, you want to kick and then
continue moving forward to administer more punishment. But
in the case of a *defensive* front kick, the goal is to stop your attacker by disrupting his balance, perhaps stunning him enough
so that you can safely run away. You won't be able to run away
if you kick the guy and lose your balance at the same time.

How does one maintain balance while executing a defensive front kick? Think about the star in Figure 2.1. What could
you do while your leg is moving forward to keep that star, or
your center of mass, from moving forward? If mass moving
forward causes your center of mass to move forward, it follows
that mass moving backward would cause your center of mass
to move backward. The key to an effective defensive front kick
is to move some mass backward while simultaneously extending your leg forward for the kick. Check out Mr. Abercrombie
in Figure 2.2. He hasn't just kicked his left leg forward, he's
arched his torso back a bit. He doesn't need to arch his torso
back much because a person's trunk accounts for just over half
a person's total mass. Remember that torque is a force times a
lever arm. Mr. Abercrombie's left leg tries to rotate him counterclockwise (as seen in the photo). His torso tries to rotate
him clockwise (as seen in the photo). Because his torso weighs
more than his leg, he doesn't have to move his torso back very
far to get the two torques to cancel each other.

Now what do you think will happen when your defensive
front kick connects with your attacker? Imagine Mr. Abercrombie kicking someone in Figure 2.2. He'll feel a force in the
direction of his right in the photo. What will keep him from
rotating clockwise and falling? He'll rotate his torso forward at
the moment of impact so as to create a counterclockwise rotation. He'll also hope he's got enough friction between his shoes

Figure 2.2
A defensive front kick.
Mr. Abercrombie forced his
torso backward at the same
time he delivered his front
kick.

and the mat to keep him from sliding backward. If he's being
pushed backward, his shoe will try to slip backward. Static friction between his shoe and the mat will hopefully hold his foot
in place.

It's possible that if you have to execute a defensive front
kick, you'll be pushed backward at the moment of impact. Instinct from many years of walking, running, stumbling, and
falling will kick in and you'll straighten up your torso. But don't
think your kick was a failure if you stumble backward and possibly fall on your butt. The drunk guy coming at you wasn't expecting to be kicked, his balance was disrupted, and, depending
on which part of his body you kicked, he might be in a decent
amount of pain. It won't take you long to get up and run away.
And if you have to stay and fight, you've already put the thought
in your assailant's head that you're not keen on being a victim.
Never underestimate the *psychological* disruption of balance
that you create in your opponent when you mount an effective
defense.

Newton's Third Law Aside

Suppose your defensive front kick was successful. You landed your kick square in the cullions[7] of your inebriated would-be assailant. You were dressed well for your evening out with friends; you were wearing solid, hard-soled shoes. The drunkard you kicked responded by falling to his knees while holding an area that seemed to grow more painful with each passing second. By the time he staggered to his feet, you were safely in the bar, having met your friends and asked the bartender to call the police so they'll have a chance of preventing another altercation outside the bar. One of your friends brought you a pint of your favorite ale and told you that you look good, except for a scuff on one of your shoes. You then told your friends the full story of what happened outside, and you let them know your shoe's scuff came from a kick you learned in your Krav Maga class and that your foot was just fine.

Okay, now for a physics question. Which felt a larger force, your shoe or your attacker's groin? Keep in mind that your attacker fell to his knees after being kicked and you ran off to the bar with no thought at all for the foot that landed your kick. Am I asking a silly question?

The correct answer is that your shoe felt exactly the same force that your attacker's groin felt.[8] Surprised? Most people unfamiliar with physics are surprised, so don't worry if you got my physics question wrong. And it's okay to be skeptical at this point and doubt my claim of equal forces. After all, you might think, your shoe was barely scuffed, yet your attacker will be sore for hours.

7. I opted for a word from Middle English that describes that place where no man likes to be hit.

8. Force is one of many *vectors* we study in physics, which means that, besides having a magnitude, it also has a *direction*. Your shoe and the attacker's groin felt the same magnitude of force, but the two forces were in opposite directions.

What I just described is a manifestation of Newton's Third Law. It says that if object #1 exerts a force on object #2, object #2 must exert a force on object #1 with the same size, but opposite in direction. You may have heard this law expressed as "for every action, there is a corresponding equal and opposite reaction," but I think that phrasing is terrible. The notion of a "reaction" suggests that the "action" came first, prompting the "reaction." With Newton's Third Law forces, there is no first force. Both forces exist simultaneously. Newton's Third Law in fact tells us that forces, like the Sith in *Star Wars,* come only in pairs. There is no such thing as a force existing all by itself. If you push down on a mat while doing push-ups, the mat has to push up on you. As soon as your hands make contact with the mat, you and the mat exert forces on each other. Don't confuse simultaneous forces with the notion of *intent.* A disputatious physicist on trial for punching someone could say, "My fist felt exactly the same amount of force from that guy's jaw as his jaw felt from my fist. Why am I the one on trial?" The physicist's claim is certainly true, but the court will be more interested in the physicist's intent than in his attempt to sidestep responsibility by acting like a pompous ass. Just because you choose to push on a mat while exercising or someone on trial chooses to throw a punch, doesn't take away from the fact that Newton's Third Law is still at work.

People struggle to believe Newton's Third Law when thinking about forces exerted by two objects with vastly different sizes. Suppose you're driving your sturdy two-ton sedan at 70 mph on an interstate and a fly smashes into your windshield. You say, "Gross!" and then flip on the wipers and squirt some wiper fluid to clear your view of what's left of the unlucky fly. Which felt a larger force, your car or the fly? Even after sharing with my physics class the explanation contained in the previous paragraph, many students claim the fly *has* to have felt more force. How could it be otherwise? Look at the fly! It was annihilated, and if it hit the car's grill instead of the windshield, you would never have known the car got hit by anything. Yet the car and the fly *did* feel the same amount of force.

The perception difficulty comes because many people confuse what's being said in Newton's Third Law with what's contained in Newton's Second Law, which I discussed in Chapter 1. The net force an object feels is equal to the object's mass multiplied by the object's acceleration. Even though the car and fly felt the same size force, the car is something like 90 million times more massive than the fly, which means the fly experienced 90 million times greater acceleration at the moment of impact compared to what the car felt. Don't confuse the size of a force with how much that force is capable of changing an object's state of motion.

So why did the same force that sent your attacker to his knees only scuff your shoes? Your attacker's groin is a vulnerable area, which means it doesn't take much force to that area to cause pain. The same goes if an attacker is hit with an eye jab or gets his ear hit with a palm heel. Your shoe, your fingers, and your palm heel can tolerate much larger forces than can an attacker's groin, eye, and ear. When you kicked your attacker, your intent was to go after a portion of his body that evolution did not prepare for being hit with large forces.

If you still aren't totally convinced by my discussion of Newton's Third Law, let me give you one more thing to think about. I've already mentioned how we are made up of atoms. But those atoms are mostly empty space. The average radius of an electron orbit[9] is about 100,000 times larger than the size of the nucleus. A boxing ring is 20 feet by 20 feet. Imagine putting a human hair in the center of the ring. The thickness of the hair would correspond to the size of the nucleus and the ropes enclosing the ring would correspond to where the electrons orbit. Every-

---

9. I'm definitely not going to go into a quantum mechanics discussion in this book. I'll just tell you that electrons don't orbit atomic nuclei in the same way the planets in our solar system orbit the sun. My comparison of the average radius of an electron orbit to the size of a nucleus is merely an order of magnitude estimate.

thing between the ropes and the hair would be empty space. Electrons whiz around in atoms with speeds of approximately 1,400 miles per second. At that speed, an electron could travel around Earth in about 18 seconds. That's fast! My point is that despite so much empty space in atoms, electrons are moving so fast that when atoms get close together, it's as if a smear of negative charge gets close to another smear of negative charge. Like charges hate being close together, meaning if the electron smears get too close together, the atoms will repel each other. The force one atom feels is equal to and in the opposite direction of the force the other atom feels. Newton's Third Law. This is why when you set a pint of beer on a table, the gravitational tug on the beer glass doesn't pull the glass and your tasty beer through the table. Because forces add together, just imagine all those atoms at the surface of your shoe getting oh so close to the atoms in your attacker's trousers. It doesn't matter whether the atoms were in your shoe or in your attacker's trousers. Put a bunch of atoms close together and they'll repel each other. This little tangent into the microscopic world may not have helped you better understand Newton's Third Law. But I have had introductory physics students with light bulbs going on over their heads when they thought about the world from the point of view of atoms. Let's move on and make use of Newton's Third Law in dissecting more Krav Maga techniques.

Vulnerable Areas

It probably goes without saying, but vulnerable areas are vulnerable for a reason. Even if you're still not convinced that a shoe and a groin feel the same force during a kick, you surely have no trouble believing that one's groin is a vulnerable area, and one certainly doesn't want to get kicked there. Getting poked in the eye, having your ear smashed, or getting jabbed in the throat aren't exactly pleasant things to experience. But what is it that makes those areas so vulnerable?

The brain has areas devoted to processing input from the five senses, and the area dedicated to visual input is the largest by far. There are so many neurons assigned to the job of processing what we see that our other senses (hearing, touch, taste, and smell) can't even compete in the "neuron number competition" when their counts are *combined*. You've probably heard or read about blind people who have acute senses of smell and hearing. Without an ability to see, they have to interact with their world using their other four senses. A blind person simply has no choice but to learn to recognize neural inputs from non-vision senses. Even the sense of touch gets more finely tuned as a blind person learns to read Braille. But it's not as if a person with normal vision can't hear what a blind person hears. It's just that our brains are so dominated by what we see that we are not always focused so acutely on each and every sound entering our ears. Setting aside pain for the moment, anything that disturbs a person's vision will likely disrupt that person's balance. If you ever get knocked on the ground by an attacker, something as simple as grabbing some dirt while on the ground and tossing the dirt in the attacker's eyes could disrupt your attacker's balance enough to allow you to escape.

Throw pain on top of messing with someone's eyes and you can subdue someone a lot bigger than you. Eye jabs don't require much force because your fingers will make contact with what's essentially a ball of tissue that contains fluid under pressure. In physics parlance, pressure is the component of force perpendicular to the area over which that force is acting, divided by that area. The fluid is the vitreous humor and it helps maintain an internal eye pressure of about 0.29 psi (or 15 mm Hg, which is a number you may hear from your optometrist). That pressure is actually a *gauge* pressure, in that it represents the pressure above normal atmospheric pressure of 14.7 psi (or 760 mm Hg). Even a low-force poke from a finger can make for a painful eye. We are taught to keep our fingers together and strike an opponent's eye with a quick jab. Fig-

Figure 2.3
Mr. Abercrombie goes from a nonassertive posture to an eye jab. Note that while he executes an eye jab, his other hand protects his face.

ure 2.3 shows Mr. Abercrombie demonstrating the proper technique for an eye jab.

If our fingers enter at an angle and miss the eye, we might be able to hit the eye with our palm heel. If you now think that an eye jab can disrupt someone's balance, imagine what an eye *gouge* could do. Just take what I wrote in this paragraph and multiply the pain by a factor of ten, at least. We are taught how to do an eye gouge, which usually follows another technique that has discombobulated our attacker.

Not only are we trained to target an attacker's eyes, we are trained to fight when our own vision is either impaired or removed altogether. Some drills have us begin with our eyes closed, only to be told to open them and respond to an attack. In some cases, we open our eyes to find there is a gun pointing at us. The reason for that type of training is to teach our brains to avoid panic if we find ourselves in a situation where our dominant means of acquiring information is suddenly turned off. Our advanced Krav Maga instruction includes an instructor putting a bag over our heads and then simulating a hostage situation. I'll delve into that fun in Chapter 10. We are trained to fight even when the world around us is hidden.

Let's move from the eyes to the ears. An ear smash is a combative[10] that is executed by keeping your fingers together and cupping your hands like they are about to hold water. Using both hands and minimizing time by thrusting your hands straight toward your attacker's ears (instead of big roundhouse movements), you simultaneously smash both ears with your cupped hands. By cupping your hands, you increase the amount of air you can inject into the attacker's ears during the smash. Pressure inside the ear rises with your hands over the ears. The increased pressure is felt in the semicircular canals, which are located in the inner ear, and which provide us with dynamic balance. Unlike static balance, which essentially lets you know if you're upside down or not, dynamic balance lets you know if you feel a net force, or if you're accelerating. If not for the semicircular canals in your ears, you could be bumped from behind and not immediately feel like you're losing your balance. Smash an attacker's ears and the fluid in those semicircular canals will move in ways that the person will definitely have his or her balance disrupted.

I doubt I need to go into much detail about why hitting an attacker in his or her throat is likely to disrupt that attacker's balance. We take in oxygen vital to our lives through our trachea. As with our eyes and ears, evolution has led to our trachea being vulnerable for a reason. If anything messes with our ability to breathe, our bodies need to send an immediate danger alarm to our brains. Imagine having your breathing tube so well protected that if it got damaged, you'd never notice. Sure, you'd be free of pain from a throat jab, but you wouldn't know that your ability to breathe was damaged until it was potentially too late to do

---

10. Warrior Krav Level 1 Certification Manual definition: "Combatives: Ballistic attacks we use to neutralize a threat."

anything about it. Keep in mind, though, that the need for sensitivity doesn't demand that a structure be vulnerable. Our brains are in a skull and our hearts are in a ribcage. Some organs are so important that we've evolved protections for them.

Other vulnerable areas, such as knees and spine, are potential targets as well. The purpose of this section is to encourage you to think about what parts of an attacker could be struck in order to disrupt his or her balance. If you punch someone's arm, it might take a dozen solid punches for that person to finally relent. But a poke to the eye can immediately put an aggressive attacker on the defensive. The other benefit to you as the defender is that an eye jab takes considerably less effort than punching someone's arm a dozen times. You'll need that saved energy to escape your attacker or, possibly, to ward off your attacker's buddies.

A Little Push Is All It Takes

Now you have a good understanding of balance, and you have seen a few ways to attack vulnerable areas. But you'll never know the skill level of your attacker. That person may also understand balance, and might be determined to go after your vulnerable areas. In the safe environment of a professional training facility, we practice various techniques without a great deal of resistance. Practice scenarios get more intense as we move through the ranks, so that we come closer to what to expect in an actual confrontation. We are also introduced to counterattacks and counters to those counterattacks. As appealing as it would be to ward off an attacker with a single technique, reality likely calls for multiple techniques.

Suppose your attacker knows something about balance, and keeps a wide stance while attacking you. In Figure 2.4, I am attacking Mr. Abercrombie. I thought I would barrel into him and knock him over, but he kept a wide stance and my initial attack wasn't successful. However, look at my stance. I know a

Figure 2.4
I attack Mr. Abercrombie by barreling into him. We both have wide stances, meaning neither of us will likely be successful pushing the other to the ground.

little physics, and I knew not to attack my potential victim with my feet together. Despite his impressive strength, Mr. Abercrombie might still need to struggle to push me down. Remember: if the winner in every confrontation was decided on strength alone, there would be no self-defense systems. We'd all be in the gym, pumping iron and, perhaps, taking some pills or injections that help bulk us up. But Mr. Abercrombie can disrupt my balance without having to exert maximum strength. The trick is to destroy the nice base I've established for myself so that my center of mass is nearly over one of my shoes. In that case, Mr. Abercrombie needs only to push my center of mass a short distance to initiate a fall. That's a much better option than trying to push my center of mass a long distance while my base is wide.

Figure 2.5
Mr. Abercrombie goes for the leg sweep.

Figure 2.6
Mr. Abercrombie is about to take me down.

The "tactical takedowns" we learn are "techniques used to subdue or restrain an attacker for control or arrest."[11] I talk about control more in Chapter 4; I want to focus on disrupting balance right now. Before Mr. Abercrombie can control me, possibly for the purpose of waiting for the authorities to arrive and arrest me, he's going to use a leg sweep to get me to fall. In Figure 2.5, Mr. Abercrombie has kicked his right leg outward while securing my right arm and holding my shirt. This is a natural movement because all he's doing, essentially, is taking a big step. Gravity will help him come down and ultimately doom me. Note that my base is still wide.

The trick to a leg sweep is to let gravity help you. Mr. Abercrombie hooks my right leg by simply stepping down and to the

11. Warrior Krav Maga Advanced Phase A Certification Manual.

Figure 2.7
I have been leg-swept.

right. As you can see in Figure 2.6, Mr. Abercrombie has reestablished his stable base. My right leg has been hooked and he has control of my right side. All that's required to get me on the floor is for Mr. Abercrombie to rotate slightly counterclockwise (as seen from above, which means as seen from the ceiling). My center of mass will then be behind my shoes and gravity can do the rest of the work in making me fall.

Figure 2.7 shows Mr. Abercrombie's rotation and my fall. He didn't have to move very much or exert a large force to initiate my fall. All he had to do was put himself in a position where he could most easily get my center of mass outside my base of support. Leg muscles are more powerful than arm muscles, so a leg sweep, followed by a leg hook that bent my right leg in a direction it likes to bend, with the hinge joint in my knee, led to the desired movement of my center of mass. Here is an example of a Krav Maga technique that exploits how we evolved. We have knee joints that bend in a given direction, so go with the bend. Note in Figure 2.7 that my left hand has slapped the floor. I did that to help spread out the force of my impact on the floor, thus lowering the pressure I experienced from the floor.[12] That is a technique I learned while studying karate. We would practice "break falls" by falling in various directions and using our hands and arms to assist in pressure reduction. Exclaiming a word like "Kiai!" allows you to release air from your lungs upon landing on the ground. That helps avoid the sensation of having your breath knocked out of yourself. Despite my efforts to lessen the intensity of my fall, Mr. Abercrombie is now primed to control my arm, which he kept control of as I fell, to inflict further damage.

Your safety and the safety of those around you are most important when dealing with an attacker. If you are able to execute a technique that allows you to maintain control of your attacker, that's great. Proper authorities may be called and the person you have under control can be arrested. But if you're struggling to keep an attacker under control, deliver a few quick combatives and vacate the danger zone. There is no shame in not playing the hero. Congratulate yourself if you get away from an attacker unharmed, and wake up the next morning ready to apply the physics you've learned in your Krav Maga class.

A Real Need to Disrupt Balance

I'm going to end this chapter with a more frightening example. It's time to get grisly and put a gun to your head. There'll be more gun examples coming in this book, so don't think the scary stuff ends with this chapter. I'll focus on a gun-to-the-head self-defense in the next chapter in the context of speed. You certainly need to be quick while performing the technique

---

12. Recall that pressure is (perpendicular component of) force divided by area.

I'm about to unveil, but my focus now is on how disrupting the balance of your attacker is such an integral part of the technique.

Look at the scene in Figure 2.8. None of us ever wants to be in that situation. But let's not pretend that there are no guns and no bad people in the world. I'm going to repeat the following comment later in this book because I don't want any reader to wind up with a broken finger. Note that I'm holding a training gun with my index finger nowhere near the trigger or inside the trigger guard. *Never practice self-defense with a training gun while your finger is inside the trigger guard.* You'll see why in a moment. Besides noticing that important safety feature in Figure 2.8, note also where Mr. Abercrombie's hands are. When dealing with guns and knives, anything you can do to reduce the time needed to execute a given technique is definitely encouraged. Mr. Abercrombie has his hands up on my left forearm. He'll need to get that gun off his head and it would take longer to get his hands on the gun if his arms were hanging by his sides. It's likely that if someone actually holds you in the manner shown in Figure 2.8, that person won't be shocked if you have your hands up instead of down.

Speed and quickness are obviously important, but I'll wait until the next chapter to break down a self-defense technique against a gun in terms of the times between the various parts of the technique. For now, look at Figure 2.9. If performed correctly, getting your hand on the gun and turning the gun can be completed easily before the attacker has time to react. Mr. Abercrombie raised his right hand, grabbed my gun, and rotated it forward. Note that Mr. Abercrombie's eyes are on my gun, where they need to be. Though you can't tell from the photo, Mr. Abercrombie has an iron grip on my gun. It's crucial that you control the weapon.

What Mr. Abercrombie does next, after having secured my gun and ensured that it no longer points toward his head, is to reach up with his left hand and secure the back of my gun and

Figure 2.8
This book just got real.

Figure 2.9
My gun is no longer pointing at Mr. Abercrombie.

my gun-holding right hand. Figure 2.10 shows this important step, and it shows that Mr. Abercrombie still has his eyes locked on my gun. Were I a real assailant, my mental balance—at the very least—would have been disrupted by this point in the action. My victim has his hands all over my gun and it's no longer pointing at his head. All of this has taken place in a time shorter than the time I needed to process what was happening in order to pull the trigger. Now, from a standpoint of disrupting my physical balance, note in Figure 2.10 that Mr. Abercrombie is dropping his center of mass. He's about to put the physics principle we learned at the commencement of this chapter to life-saving use. How to disrupt my physical balance? Get my center of mass outside my base of support.

In Figure 2.11, Mr. Abercrombie has not only dropped his center of mass, he's taken my gun—which would probably have broken my finger if it weren't a training exercise—by forcing

Figure 2.10
Mr. Abercrombie has secured my gun with both hands and has dropped his center of mass.

Figure 2.11
Mr. Abercrombie has my gun and has maintained his stability. My balance isn't doing so well.

both his arms forward. But he's only going to maintain his stability if his center of mass doesn't move away from his base of support. That means that if he pushes some of his mass forward, he has to push other parts of his mass backward. And that's just what he does. As his arms thrust forward, he simultaneously thrusts his butt backward. That keeps him stable, but completely wrecks my stability. With my arms still around Mr. Abercrombie, his butt moving backward means that my center of mass is now in front of my base of support. All he has to do is step forward and I'll fall forward. While falling forward Mr. Abercrombie would take advantage of my lost balance by jabbing an elbow back toward my oncoming face. You already have the key piece of physics highlighted in this chapter. The reason this self-defense against a gun to the head is so effective is simple physics. You obviously have to be skilled and fast with your technique in a real situation, but the beauty of the self-defense technique is that all Mr. Abercrombie had to do was re-orient his body in such a way as to make me lose my balance in exactly the same way he lost his balance in Figure 2.1, by getting my center of mass out past my base of support. For all the necessary aggression and fast movement in a Krav Maga technique, there is usually a simple physics idea at the heart of it.

The Need for Speed

One of the questions in my head while I'm training in Krav Maga class is, "Am I fast enough to do this technique in real life?" That is surely a question in the minds of many martial arts students. If someone attacked me for real, could I react quickly enough to defend myself? We gain confidence in training where we repeat techniques many times so that they become second nature. We do improve on speed as we move through the ranks, but the worrying question in my head hasn't quite left me yet. What I hope to do in this chapter is give you an understanding of speed and let you know that you don't have to be "fast as lightning" to successfully defend yourself.

What Is Speed?

What do you think of when you hear the word "speed?" A race car? Maybe a drug that helps you stay awake? I find many people equate "speed" with "fast," which is why my chapter title suggests the need to go faster. But physicists use the word "speed" in a precise way, and it may surprise you that we don't use "velocity" as a synonym for "speed."

Speed is a measure of the rate at which distance is traversed in a given time. There is no direction associated with speed. If you drove 50 miles at a constant speed, and it took you an hour to drive that distance, your speed was 50 miles per hour (50 mph). It doesn't matter if you drove north, south, east, west, or some other direction. It also doesn't matter if you drove 25 laps on a two-mile racetrack and finished your 50 miles of driving at the same place where you started. Speed is what your speedometer shows you (more on that in a moment). The "distance is rate times time" idea that you may have learned in school is what allowed you to find speed (or rate) if you knew the distance (50 miles) and the time (1 hour).

One problem with this simplistic "distance is rate times time" idea is that it's not immediately helpful if your speed changes. Suppose you drove a total distance of 50 miles, but you stopped along the way to check a map. Suppose further that part of your trip was on a piece of interstate on which you drove 70 mph and another part of your trip was through a neighborhood in which you drove 25 mph. The 50-mile trip could still have taken one hour, but your car clearly had several speeds during that hour. What then is 50 mph? That is your *average speed*. Though you would have angered drivers on the interstate and scared people in the neighborhood had you actually driven 50 mph along the same route, you would have gotten to the same ending point in an hour, just as you did with regular driving.

So, the problem with "distance is rate times time" is that it is useful only as long as the rate (or speed) is *constant* during the time the distance is traversed. But on the aforementioned hypothetical trip, your speedometer had many different readings during your trip. What your speedometer shows you is your *instantaneous speed*. That is, "distance is rate times time," but with a very small time interval. That is the realm of calculus, which we won't be discussing here.[1] Despite the subtleties, the key point is that we can calculate speed by measuring the distance something moves over a given amount of time.

You should have a few common speeds in the back of your mind. The idea isn't to have precise numbers memorized, but you should at least have a general sense of how fast various

---

1. Even though we're not discussing it, this imaginary trip illustrates why we *need* calculus. Things move, but things rarely move at the same speed while moving. Starting and stopping imply that speed must either change *from* zero or change *to* zero. Even predicting the motion of your car during such tasks as backing out of your driveway and driving to work requires calculus.

things move. For example, if someone told you that a typical car speed on the interstate is 7 mph, you'd know that's piffle because interstate speed limits are in the 55–70 mph range. How fast do you think a person can run? A world-class athlete can run the 100-meter sprint in something like ten seconds. That's an average speed of 10 m/s or about 22 mph. People in great shape can therefore run close to 20 mph, but only for short distances before getting tired.

You've probably heard that running a mile in four minutes is pretty good. In fact, it's so good that no human being had been timed doing it until Roger Bannister beat that mark by 0.6 seconds in 1954. A person that can run a four-minute mile averages 0.25 miles per minute, which is the same as 15 mph—that's school zone speed limit territory. In the context of our Krav Maga science, keep in mind that if you have to run from an attacker, a burst of speed close to 20 mph may be necessary to get away quickly, but sustaining a speed of 15 mph may be what's required in a foot chase that lasts a minute or two. Those speeds may be easy to visualize from behind the wheel of a car, but maybe you've not thought much about running speed. A useful conversion is 15 mph is exactly the same as 22 ft/s. If the length of two strides is close to one's height, and two running strides are perhaps a little greater than one's height, a person 5½ feet tall requires four strides to cover 22 feet. That's easy to check. If you don't have a radar gun handy, and if you don't have immediate access to a track, have someone time you as you run four (or eight) strides down a hallway. Can you make those four (or eight) strides in one (or two) seconds? That will give you at least a very rough idea of how well you could escape someone running after you. If someone with Usain Bolt–like speed decides to attack you, you need to forget about running and get to more self-defense.

Have you ever thought about sound and light speeds? Sound speed depends on various atmospheric properties, such as air temperature, but for our purposes here a reasonable estimate is

770 mph. Light speed is *much* greater. In air, light travels about 186,000 miles in one second, which is roughly 870,000 times greater than sound speed. Light travels so fast that we often take its speed to be infinite when distances aren't too great. It takes sound around 4.7 s to travel a mile, but it takes light 5 millionths of a second to travel a mile. That time for light to travel a mile is about 1/50,000 of the time it takes to blink. That's why, when you were growing up, parents and/or teachers taught you to determine the distance you are from a lightning strike by counting the seconds between seeing the lightning and hearing the thunder (which was caused by the lightning strike). The rule of thumb is to divide the number of seconds you counted by five and that gives you the distance in miles you are from the lightning strike. As I mentioned, sound takes about 4.7 s to travel a mile, but the time needed for the light to reach you is so short that we can ignore it.

I discuss sound and light speeds because you may witness a crime or other heinous behavior from some distance. You may see objects collide, like a baseball bat hitting a person or two cars hitting each other, and hear the sound after you see the collision. Anyone who's sat in the outfield bleachers at a baseball game knows this phenomenon well. The crack of the bat hitting the ball is heard after the bat is seen hitting the ball. If you ever do witness a crime from a distance, you may be too frazzled to think about anything like sound and light speeds. But if you're calm and, say, happen to notice that you hear a collision about a second after you saw the collision, you can estimate that you are about a fifth of a mile (1,056 ft) from the collision. That distance is more than three football fields set end to end, so you probably have good eyes, binoculars, or a high vantage point for witnessing the crime. Keep in mind that knowledge is power, and every little detail you can put together in your mind helps you make a good decision.

So what's "velocity" and why is it different from speed? Velocity carries *direction* as well as magnitude. In other words, toss in your GPS reading with your speedometer reading, and then you'll know if you're driving 50 mph to the north or northeast or some other direction. We have to be careful not to think of speed as simply the magnitude of velocity because the latter carries with it a notion of *displacement,* which also carries direction information, whereas the former requires distance traveled (as in what's on your car's trip meter). In the racetrack example I gave earlier, the car started and ended at the same place, so the displacement is zero, meaning the *average velocity* is also zero. Confused? All that means is that a car parked at the starting (and ending) point would be at the same place after one hour as the car that was driving around the track.

You already have an intuitive understanding of speed. When I use the word "speed," I'm describing the instantaneous rate at which distance is being traversed in time. If I need to include a direction an object is moving, I'll use "velocity."

There are many popular units for speed, such as miles per hour (mph), kilometers per hour (kph), feet per second (ft/s), and meters per second (m/s). Any distance unit divided by any time unit will make a unit of speed. We choose units appropriate to a given situation. The speed of a fist moving through the air might best be described with a distance unit comparable to the length of an arm and a time unit comparable to how long a punch lasts. Fist speed in ft/s or m/s works well. Many in the United States are familiar with mph, and that unit might work well, too, even for a moving fist, because some people may have a better feeling for mph than for ft/s. I'll use units popular in the United States, but it's easy to convert from one unit of measurement to another.[2]

---

2. For the popular speed units I mentioned, 10 mph $\simeq$ 16 kph $\simeq$ 15 ft/s $\simeq$ 4.5 m/s. I've kept just two significant digits. Nothing we do in this book requires too much numerical precision.

Hammer Fist

I'll go into more detail on punching in the next section. I want to get our Krav Maga science training started with a single punch—a hammer fist. We learn how to execute a hammer fist early on because it's a simple, medium-range combative. Suppose you catch someone coming up behind you or from the side. You've determined that you need to be preemptive because the person coming at you means to hurt you. You could, in fact, have been engaging one attacker and noticed his mate coming up behind you to help in the attack. You need to move with speed so that you don't have to fight two people at once.

My instructor, Mr. Abercrombie, is training me in Figure 3.1. I've got my combat gloves on for safety and my hands are up. Mr. Abercrombie has come up behind me on my left side and he's holding a body shield so that I may practice a hammer fist. We are taught to see our target before striking, which is why I'm glancing over my left shoulder. Mr. Abercrombie has a wide stance, which means he's established a good base for stability and he's prepared for me to strike the pad.

Remember the theme here is speed. I need to strike my attacker before that individual inflicts harm upon me. If I'm currently fighting someone in front of me, my strike to that person's mate had better be quick. Look at Figure 3.2. I've delivered my hammer fist to the pad in 0.5 s from the time seen in Figure 3.1. That half second is quick enough to strike my stealthy assailant before that person's reaction time kicks in for an adequate defense. I must be mindful of the fact, however, that the person coming up behind me may be well trained in fighting and may already have his guard up, anticipating that I might strike before he reaches me. There are no guarantees in self-defense. But let's suppose I'm lucky enough not to have a professional MMA fighter coming up behind me, or my instructor, for that matter. Suppose I landed my hammer fist. I managed to hit my target with a fist moving at 15.5 mph. I'm not the

Figure 3.1
I look over my left shoulder
and see danger coming.

fastest and strongest person in my class, but that's enough speed to do some damage. At the very least, I'll stun the person enough that I can resume fighting my original assailant or, if the person sneaking up behind was alone, I can continue administering combatives until I feel I can evacuate the danger zone.

I can develop better speed by working on my core strength and fast-twitch muscle system. Repetition while training will also improve my speed. But there is something else I can do better. I put Figure 3.2 in this book for a reason. I made a big mistake. Can you see it? My right hand dropped while I delivered my hammer fist. I dropped my guard. You can see Mr. Ab-

Figure 3.2
I've delivered my hammer fist, but dropped my guard.

ercrombie's well-trained eyes staring at my lowered right hand. Speed may have been in my favor as I delivered a half-second hammer fist, but my attacker (or attackers) can make use of speed, too. Had I been previously engaged with someone in front of me, my hammer fist on that person's mate has left my head wide open to attack. I might take a hit to the right side of my face without even knowing what happened. Had the person behind me been alone—and prepared for my combative—my hammer fist could have been evaded and I would have been vulnerable to a strike to my head. I would have been the one hoping for fast enough reaction time to cover my head before getting hit. Don't ever think there aren't mistakes in training, even after a few years in Krav Maga class. I made a mistake, and my instructor was good enough to spot it and call me on it. I cover my blind side better now. We can't violate the laws of physics, and we shouldn't ignore the sage advice of our instructors. I recommend doing what I've done in this book. Film yourself. Go frame by frame using *Tracker* or whatever software you like. Analyzing video is a great way to spot mistakes in your technique. Show your instructor your analysis, as I've done, as you'll likely receive more sage advice, as he or she is able to spot errors that you might miss.

Punches

Krav Maga is about doing whatever it takes to defend yourself. Training requires getting in shape, which includes developing fast-twitch muscles capable of releasing energy quickly. Developing good slow-twitch muscles is important, too, but more so if you are interested in running a marathon, where endurance is key. Energetic bursts provided by fast-twitch muscle fibers may be used to disrupt an attacker's balance, counterattack quickly, and then speedily vacate the danger zone. Fast-twitch muscles may be honed by vigorous movements with medicine balls and box jumping. Using heavy weights and fast lifts also helps develop fast-twitch muscles.

Though fast-twitch muscles can unleash fury in just a few seconds, they fatigue easily, which is why they aren't helpful for running marathons. The rapid onset of fatigue is why some fights don't last long; the combatants get tired quickly. Your greatest weapon in an attack could be your ability to counterattack quickly and not get fatigued before your attacker tires. Your exercise regimen could provide you with a small window of time during which you can escape a more fatigued attacker.

Let's start with a relatively simple fight technique: punching. Every person reading this book will have punched something in his or her past, like a punching bag. If we don't have a weapon in our hand to swing or fire, we use our fist as a weapon.

We make a fist so as to concentrate our punching force onto a small area because we are likely to do more damage to an attacker with a punch than with a slap. Figure 3.3 shows my right fist. The rectangle in the photo encloses the part of my fist that should be used to hit a target. That contact area is about 2.3 in^2 (15 cm^2), which is roughly three times the area of a postage stamp. Anyone who has ever laid on a bed of nails knows that *pressure* is more important than force. Pressure is the force perpendicular to an area divided by that area. No one will lie on a bed with one nail because a good chunk of the person's weight will be concentrated on the small area of the point of the nail and it will puncture skin. Put a few hundred nails on the bed and the person's weight is distributed over a much larger area. The same idea works with a fist. A large striking force using a small area results in a large pressure on the target, possibly capable of puncturing skin. A pressure of about 100 psi (pounds per square inch) is required to puncture skin, which means if I hit with a force of approximately 230 pounds using the rectangular area shown in Figure 3.3, my punch could puncture skin. I wouldn't need nearly that much force if my punching area shrank to one or two knuckles. The situation is a little more complicated than simply hitting with 230 pounds using the area shown in Figure 3.3. The skin getting hit is likely to pull the surrounding skin down and the tissues just under the skin's surface will compress while the skin is hit. Putting those and other complications (like hitting skin close to bone, such as the cheek) aside, the 230-pound force I gave for my punching area isn't a bad estimate.

My outstretched right arm is shown in Figure 3.4. The arrow in the photo shows roughly the location of the line of punching force. As long as I punch with that force line along my radius (I'm speaking of the bone in the forearm called the radius), the chance of rotating my hand is low. If my wrist is bent, however, the line of force would not be lined up with one of my forearm's long bones (radius is close to the thumb; ulna is close to the

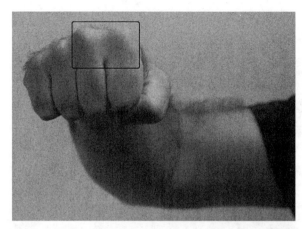

Figure 3.3
My right fist with a rectangle showing the part of the fist that should hit a target.

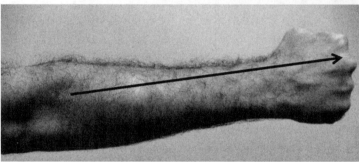

Figure 3.4
My outstretched right arm. The arrow shows a line of force along the approximate location of my radius.

pinkie finger). A torque results when my wrist is bent, which may lead to injury. Never forget that you can injure yourself with poor technique, and your injury might have nothing to do with what your attacker did.

Jab Cross

There is a lot more to a punch than flailing your arms. A standard boxer's stance is shown in Figure 3.5. As discussed in the previous chapter, having your legs at least shoulder width apart helps with stability. My Krav Maga instructor, Mr. Abercrombie, is right-handed, so he has his left fist farther from his body than his right fist. His left hand is for a jab; his right hand is for a cross. A jab is a speed punch, but less powerful than a cross. There is a slight body rotation (clockwise as seen above Mr. Abercrombie) for a jab, plus a little forward movement. Both the

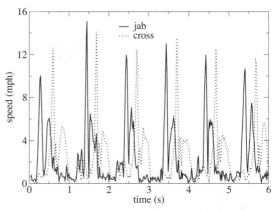

Figure 3.5
Mr. Abercrombie in a
boxer's stance.

Figure 3.6
A plot of hand speed versus
time for both a jab and a
cross.

rotation and forward movement serve to transfer energy from the main part of the body to the fist. The energy associated with a fist in motion is called *kinetic energy*. Because Mr. Abercrombie's jab hand is closer to his potential attacker, it has less distance to travel than his cross hand. A jab often serves as a quick punch to stun an opponent. The jab sets up what follows, which is a powerful cross.

A successful jab that puts an attacker off his or her balance might cause that attacker's guard to come down, thus opening up a vulnerable area for a cross. A cross follows a jab with very little time delay. Because the cross originates from the back hand, that hand travels a greater distance than the jab hand. The entire body rotates (counterclockwise as seen above Mr. Abercrombie) and transfers a large amount of energy to the moving cross hand. In physics we speak of doing *work* on something to alter its kinetic energy. Work is, loosely put, a force multiplied by a distance. The greater distance the cross hand moves compared to the jab hand means the body is capable of exerting more work on the cross hand, thus giving it more kinetic energy. A right-handed person like Mr. Abercrombie opts to use his dominant hand for the more powerful punch.

Check out Figure 3.6. I filmed Mr. Abercrombie executing a sequence of six jab-cross combinations. With tape on his fists that allowed me to mark their locations in each frame of the video, I could determine his fists' positions, and from those, his fists' speeds. There are a couple of things to note immediately. His maximum jab and cross speeds are similar. Mr. Abercrombie was not hitting a pad; he was punching the air. To avoid hyperextending his arms, he did not punch with maximum force. We hit pads a lot harder than we do the air, but it's harder to record fist position when the fist impacts a pad. A complete jab-cross combination begins with hands guarding the face, followed by a jab, then a cross, and ending with hands back guarding the face. Note that Mr. Abercrombie executes a complete jab-cross combination in less than a second.

Think about what happens when you throw a punch. Your fist begins at zero speed. You punch, and when your arm is fully outstretched, your fist speed has to hit zero again. If not, your fist would have to keep moving. Somewhere between those two zeroes has to be a maximum speed. It won't surprise you to learn that the maximum speed is roughly in the middle of the distance traversed by a fist. Think about that when contemplating the distance to your punch target. If you plan for a distance that's equivalent to the length of your outstretched arm, your punch won't do much damage.

To better understand what is happening with a jab-cross combination, see Figure 3.7, which focuses on Mr. Abercrombie's second, and fastest, of the six combinations. The jab is the solid curve; the cross is the dotted curve. Each punch consists of a rapid increase in speed, followed by a precipitous drop in speed, and then another rise and drop in speed. In each case, the second rise is not as large as the first, and the drop in speed

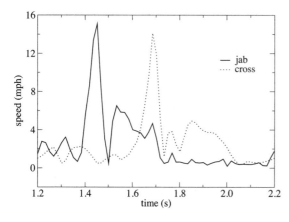

Figure 3.7
A single sequence of jab, followed by cross.

happens over a longer time. As the graph shows, Mr. Abercrombie's maximum jab and cross speeds are in the range of 14–15 mph, with his fists reaching maximum speed while his arms are extended roughly halfway. After the big punch peak, the speed drops to essentially zero as his arms have been fully extended. Pulling his fists back reverses the punches, which means another speed maximum halfway back, followed by his fists coming to rest near his face. His fists move only about half the speed they did on the way out, which means they take longer to return to his face.

Even though Mr. Abercrombie is punching at the air, appreciate how quickly his punches are delivered. It took him 0.08 s to get his jab fist to maximum speed. That doesn't give an attacker much time to respond. If the jab has done its job and his attacker is momentarily off-balance, Mr. Abercrombie's cross will land about 0.23 s after his jab hit. Imagine you've just been stunned by a quick jab, and then you've got less than a quarter of a second to get your bearings before a powerful cross tags you. If executed well, a jab-cross combination can be very effective.

Blitz

There are very few techniques in this book that don't benefit from speed. When faced with a harrowing situation, techniques executed rapidly help disrupt an attacker's balance, administer damage to an attacker so as to allow escape, and, most important, don't allow an attacker time to inflict damage on you. The quicker our reactions, the greater the temporal windows in which we can defend ourselves. Speed may be improved by acting with controlled aggression. We are taught that an aggressiveness drill "instills a survival mindset and develops our killer instinct."[3] As macabre as that might seem, keep in mind that none of us can predict the future with great accuracy. We have no idea if an attacker wants to rob us or has intentions more pernicious. We don't leave our houses in the morning expecting our lives to be at risk, but if we are attacked, we have to fight back, and we have to do so without knowing our assailant's intent.

Imagine a situation in which you have been surprised by someone's shove. Maybe you were walking down a quiet sidewalk and some guy walked out behind a van and pushed you into a wall. You're certainly startled and scared, and you don't have a great escape route that allows you to run away. Your assailant shows no sign that his push was an accident and he is ready to fight you. There is no way for you to know what is his intent. You must defend yourself. Aside from the fact that your attacker knows nothing about your capabilities, your most important asset at that moment is your ability to avoid the mindset of a victim. It's quite possible that your attacker will not be expecting an immediate reaction to his shove. You may only have a small window of time in which to take maximum advantage of the aforementioned important asset.

In Figure 3.8, Mr. Abercrombie shows a posture that you may assume immediately after being pushed into a wall and facing your attacker. His feet are more than shoulder width apart to slightly lower his center of gravity and increase his balance. His arms are up in a defensive posture. The fact that his hands are not curled into fists may suggest to his attacker that he doesn't want to fight. Any added fraction of a second that contributes to an attacker's hesitation is immensely valuable. What Mr. Abercrombie is going to do next is blitz his attacker, using speed to gain the upper hand.

3. Warrior Krav Advanced Phase A Certification Manual.

To avoid being faked out by an opponent's eye and head movements, we are taught to focus on the opponent's center of mass. A net, external force on a person causes that person's center of mass to accelerate. Eyes and head can move with little to no movement in a person's center of mass. That's why we are susceptible to being faked out if we focus on our opponent's eyes. If we watch the center of mass, however, we detect changes in motion that reflect the movement of our opponent. The blitz certainly can't violate the laws of physics, but it can make use of those laws to great advantage. Newton's Second Law tells us that a *net,* external force is required to accelerate an object's center of mass. But if an object has forces pulling on it in opposite directions, the net force could be zero and the center of mass won't accelerate. You know what will happen if Mr. Abercrombie pulls his leading leg backward, which is his left in Figure 3.8. If his left foot pushes the mat forward so as to make his left leg move backward, then his center of mass will accelerate backward—if no other part of his body moves. What could Mr. Abercrombie do to keep his center of mass where it is while pulling his left leg backward? Some part of his body will have to move forward. This is why the blitz is so effective. If the first thing an attacker notices is his victim's leg moving backward, he might think his victim is retreating. That backward movement could actually cause him to move forward, toward his attacker. But that backward movement is a lure to help Mr. Abercrombie quickly close the distance between himself and his attacker.

Figure 3.9 shows Mr. Abercrombie has pulled his left leg backward, while at the same time lifted his right leg upward and slightly forward, and his torso is leaning forward. His arm movements won't rotate his body because he pulled his right arm back as his left arm moved forward. But note that his torso wants to rotate clockwise (as seen from above), but his legs want to rotate counterclockwise (as seen from above). The two rotations cancel due to conservation of angular momentum.

There is no net, external torque on Mr. Abercrombie. That's why his head and the central, vertical part of his body haven't rotated. Look at the line on the floor where two mats come together in Figure 3.8 and Figure 3.9. Mr. Abercrombie's center of mass was roughly over that line and it hardly moved when he started his blitz. That's the power of the laws of physics. They can't be broken, but they can be satisfied in different ways.

Now imagine yourself as Mr. Abercrombie's attacker. You've shoved and bounced him off a wall and watched as he steadied himself, looked at you, and had his hands unclenched as if he didn't want to fight. You then see his left leg move backward, and your first thought is that he could be retreating, perhaps in fear. Maybe your other victims have retreated in fear, and you expect your next victim to do the same. But you messed with the wrong dude. Just 0.5 s elapsed between Figure 3.8 and Figure 3.9. All it took was half a second for your potential victim, who appeared to be retreating, now to appear to be coming at

Figure 3.8
Mr. Abercrombie is ready to fight his attacker.

Figure 3.9
Mr. Abercrombie has begun his blitz.

you. If you were lured into thinking it was retreat, even if only for a fraction of a second, you are now fighting your own reaction time and being stunned by Mr. Abercrombie's sudden aggressive blitz. And Mr. Abercrombie is by no means finished making great use of physics. The forward lean his body has in Figure 3.9 means that gravity will help his next movement. If all he did in Figure 3.9 was stop in that position, he would fall forward because his center of mass, which hasn't moved much from Figure 3.8, is no longer over a stable base, it's over an open area and his left foot is now the pivot point. Gravity exerts a torque that causes Mr. Abercrombie to rotate forward. That's just the direction he wants to go. And if his attacker actually moved forward because of an erroneous belief that he was retreating, Mr. Abercrombie will be able to deliver punishment even faster.

It's now time to deliver that punishment. Figure 3.10 shows Mr. Abercrombie moved forward a couple of feet. The stored potential energy in his cocked right arm in Figure 3.10 has been released into kinetic energy. He gets added help on his right-handed punch (or eye stab) by reversing the two rotations that led to Figure 3.9. His left arm has been pulled back to his face, guarding against a blow to his head. His right leg has been kicked backward as his left leg launched him forward. It took 0.47 s to go from Figure 3.9 to Figure 3.10. That means that Mr. Abercrombie went from an upright, stable, possibly nonassertive position to punching his attacker's face in slightly less than one second. Even if the attacker wasn't fooled by the faux retreat, being able to react to Mr. Abercrombie's blitz in less than a second takes a gifted fighter.

Don't think that Figure 3.10 represents the end of Mr. Abercrombie's attack. His body is still moving forward. Though I won't show more photos here, you can tell in Figure 3.10 that his body is indeed still moving forward. One punch isn't enough. Mr. Abercrombie wants to engage the attacker up close, push him back, punch him again, and overwhelm him with force. The

Figure 3.10
The blitz connected.

blitz is effective because of speed, but note that I mentioned several other aspects of physics, such as center of mass, gravity, torque, and angular momentum. As I noted in the Preface, all laws of physics play a part in the techniques I discuss. I will only highlight one per chapter, but I mention the others at play. I'll discuss energy and rotation in more detail in later chapters. I prefer to throw you into the action and then throw some physics at you. The blitz uses speed, and what comes next absolutely requires speed.

What If You Have a Gun to Your Head?

No, I'm not being figurative with my section title's question. I mean it literally: What if someone has a gun pointing at your head? What you see in Figure 3.11 is me pointing a practice handgun at the head of Mr. Abercrombie. He is about to demonstrate a Krav Maga technique for defending oneself against an assailant who points a gun right at the head of a would-be crime victim.

Figure 3.11
A staged scenario of having a gun pointed to your head. The gun is made of plastic and used for training purposes only.

Does what you see in Figure 3.11 look a bit scary? I'm certainly not crazy about seeing a gun pointed at anyone's head. Using practice weapons in a safe environment, I train with several others in Krav Maga, and part of our training involves escaping the frightening situation mimicked in the figure. I frankly have no idea how well I would do if I was ever unfortunate enough to find myself with a loaded gun pointed at my head. I hope that training will give me an automatic response so that I'll at least be able to fight back.

What I wish to focus on here is the physics behind what Mr. Abercrombie is about to do next. His main goal is to get the gun's barrel out of line with his head—and do it quickly! What about moving his entire body to the side? Too much mass. It will take too long. He will move his arms, which are much less massive than his whole body. The best athletes have been measured to have a reaction time of not less than about 0.2 s. During those measurements, however, athletes knew what was coming and how they were to react. For most of us in a car, we have a reaction time of at least a full second. The study of human reaction time involves neurophysiology, biomechanics,

and a host of other sciences. First the eye (or some other sense organ) detects something, then the brain processes a signal, and then the brain tells the body to do something. That's obviously highly simplistic, but if an attacker with a gun isn't expecting resistance, assuming the attacker's reaction time to be a full second isn't likely to be that far off. Note that I use reaction time to mean the time from Mr. Abercrombie initiating his defense, which may be imperceptible to me if I'm staring at his face and perhaps yelling at him, to the time when I fire the gun in response to my brain finally processing that he is fighting back and getting a signal to my hand to fire.

Let's now turn to what Mr. Abercrombie needs to do. Figure 3.12 shows the first two sequences of the defense. It took Mr. Abercrombie just 0.8 s from the first wiggle his hand made that showed he was initiating a defense to getting his hands on the gun, as shown in the left photo. You may think that 0.8 s is cutting it close, given that the attacker may be able to react in 1.0 s—but realistically, you may consider Mr. Abercrombie's move to have taken only half that long. Having witnessed it firsthand, I can tell you that I would not have distinguished his

Figure 3.12
Getting the gun's barrel off your head is the first step.

Figure 3.13
Maintain control of the
assailant's hand while
removing the gun.

initial hand movements from normal motion. The first half of
that 0.8 s probably passed before I even noticed that he was trying to defend himself. So, even if I, the attacker, can react in
0.5 s, he still has his hands on the gun before I can fire. The photo
on the right in Figure 3.12 shows that his head is out of the gun
barrel's firing line. The time between the left photo and the right
photo is just 0.03 s. Even if I can get a shot off, it will sail over

Figure 3.14
The attacker has been
disarmed.

his head. That could, of course, still be a dangerous scenario if there are people behind him and/or elevated and behind him.

Note in Figure 3.11 that I kept my finger off the trigger and along the side of the gun. That's very important while practicing. Watch what happens next in Figure 3.13.

Mr. Abercrombie rotated my hand and gun clockwise (as seen by him) and away from him toward his right. He kept rotating such that if my finger were on the trigger, he would have used enough torque to have broken it. My price for messing with the wrong person is a spiral fracture. That's part of the technique. Distract the assailant with something to think about (like pain from a broken finger) while the tables get turned. I am officially disarmed in the photo on the left, and that took Mr. Abercrombie just 0.16 s after he had the gun's barrel clear of his head. The photo on the right is 0.1 s after the left photo. Note that Mr. Abercrombie not only has control of the gun with the barrel pointed away from him and toward me, but he also still maintains control of my hand. That's important because a real attacker is likely to be fighting back at this point.

Now we get to the end and I, as the dumbfounded faux attacker, have realized that I messed with the wrong man. Fig-

ure 3.14 shows that I'm left with an empty hand while Mr. Abercrombie has the gun in what's called a retention stance.

Just 0.16 s after the right photo in Figure 3.13, Mr. Abercrombie released my hand. He then cleared away and secured the gun to his side. I'm left with an open hand. The entire technique from the first photo I showed in Figure 3.11 with Mr. Abercrombie beginning to move his arms to the final photo on the right in Figure 3.14 took just under 2 s. That was all the time needed for my Krav Maga instructor to react to having a gun pointed at his head to holding the gun himself while at a safe distance from me.

Training for such encounters is incredibly important. Knowing a little physics associated with the technique can only help. Quickness is associated with acceleration, which is a measure of how fast velocity can change. Because Newton taught us that a force is required to change velocity, and that the amount of force needed for a given acceleration is proportional to the object's mass, we move our arms toward the gun instead of moving our entire body away. The trick is to beat the physiology of human reaction time with some good physics.

Take Control

Control. None of us likes to think we are not in control of our lives. Control is the very thing we lose if an attacker gets the upper hand on us. We lose control over our possessions if we are robbed. Losing our autonomy if kidnapped is a truly harrowing scenario to imagine. Victims who survived violent attacks may have to spend the rest of their lives with limited control of their bodies because of injuries they suffered while being attacked.

Part of our Krav Maga class is spent learning "defensive tactics," which are "methods we use to deflect, intercept, or evade an oncoming attack."[1] We must learn to recognize danger quickly and react to that danger as fast as possible. Getting out of harm's way is sometimes the best thing to do. But if we are engaged with an assailant, regaining control from that person is essential. Some confrontations may end abruptly when an attacker feels he or she has temporarily lost control. That's not what the attacker had in mind. Standing up for yourself may be enough to dissuade an attacker from making a second attempt to take control from you. Turning the tables on an attacker and regaining control is what I'll focus on in this chapter.

Head and Side Control

What percentage of your body weight is your head? For most adults, the head represents a little over 8% of their total body weight. That means about one-twelfth of your weight is sitting above your shoulders. Recall that changing an object's state of

1. Warrior Krav Level 1 Certification Manual.

motion requires a force. But the more mass an object has, the more force is required to achieve the same change of motion state. It's easier to move a tennis ball around than a bowling ball, right? And you don't even need to know that a 16-lb bowling ball is 128 times more massive than a 2-oz tennis ball.

There are a couple of reasons why controlling and moving an attacker's head is to your advantage. The small mass reason was stated in the previous paragraph. Another reason is taught to us in Krav Maga class, specifically, "where the head goes, the body follows."[2] Think about that. If someone is able to control your head and pull it in a given direction, you feel obligated to allow the rest of your body to follow your head. Despite the sensation that some part of us looks out through our eyes, we don't have bodies, we *are* bodies. Our head houses our all-important brain, sensitive bones in our ears that provide us with a sense of balance, and our head is connected to our neck, which contains essential arteries, veins, and trachea. Nobody has to tell you that if you lose your head—literally—you will be in deep trouble. It is a protection mechanism to force our lower bodies to follow our heads if our heads are pulled in a given direction.

One of the first techniques we learn is "head control." The left photo in Figure 4.1 shows my head being controlled by my Krav Maga instructor, Mr. Abercrombie. Three points of control are essential because, as with a tripod or a lineman in a three-point stance, having three points braced allows for much more control and balance than having just two points braced. Mr. Abercrombie has my head squeezed between his forearms. Those two bracing points are combined with having my head clinched against his chest. Mr. Abercrombie's fingers are locked behind my head. I could flail my arms at him, but he could easily sidestep my efforts. Once he's controlled my head, he's got the upper hand. My head houses my eyes, ears, nose, and mouth (most of my sense organs), and if Mr. Abercrombie de-

2. Warrior Krav Level 1 Certification Manual.

cides to move my head around, my body will follow as I try my best to maintain some semblance of balance. Besides losing balance, I'm likely to feel stressed. If I'm lucky enough to get separated from Mr. Abercrombie, as the attacker I may feel that I've targeted the wrong person. Having the tables turned on me disoriented me enough to get me to think about running away.

Figure 4.1
Head control is on the left and side control is on the right.

What if you have an attacker in head control and that person reaches for a knife or gun? Flailing arms may not reach you, but they'll easily reach a weapon in your attacker's pockets or waistband. What if your attacker is trained well enough to offer a counterattack? It's not as if your attacker will stay calm while you move his or her head around. Head control is not meant to be a fight-ending technique. If you manage to gain control of your attacker by putting him or her in head control, you need to do something else to ensure that when you release your attacker, that person won't be an immediate threat to you. You will need to unleash "combatives," which are "ballistic attacks we use to neutralize a threat."[3]

3. Warrior Krav Level 1 Certification Manual.

The right photo in Figure 4.1 shows "side control," during which the attacker's head is held down with a hand or forearm while one of the opponent's arms is locked with either an underhook (shown) or an overhook (not shown). Note there are three points of contact in the side control Mr. Abercrombie has me in: under my arm, my shoulder, and the back of my neck. Because Mr. Abercrombie has hooked my left arm, he has his left leg back. That keeps his body out of the way in case my right arm either flails at his body or acquires a knife from my pocket or waistband. Several of the defensive tactics in Krav Maga are designed to flow into side control so that combatives may be administered quickly and violently with the goal of neutralizing a threat.

There is another reason Mr. Abercrombie's left leg is back. It is now in perfect position to drive his left knee into my face. His left knee will move along a much greater distance than his right knee could move. By using his leg muscles to act with a force over a large distance, Mr. Abercrombie's knee will have work done on it so as to increase its kinetic energy. That translates into a high speed for his knee when it collides with my face. Not good for me. Mr. Abercrombie might elect to swing his left leg down while rotating his hips clockwise (as seen from above) and deliver a roundhouse kick. Once I've been kicked good enough to disrupt my balance, disorient me, and cause me stress, Mr. Abercrombie can release his left arm from my neck and then deliver punches and other combatives to sensitive areas within my head, such as my temple, eyes, ears, and/or top of my spine.

All of the aforementioned combatives are meant to neutralize the threat as fast as possible. Once I'm beaten up, Mr. Abercrombie can back away while swiveling his head around "to look for additional attackers, escape routes, and weapons of opportunity."[4] Never celebrate early. You may have gotten the upper

4. Warrior Krav Level 1 Certification Manual.

hand on your assailant, and you may have even delivered enough punishment so that person won't be bothering you anytime soon, but your attacker's friends could always be lurking in the shadows. Note exits so that you can get away quickly if you are chased. If you see a knife on the ground, it might be worth grabbing so that you have a weapon in case you need to fight again. Even getting a handful of dirt may be useful. Tossing it in the eyes of your assailant's friend could leave him or her wiping their eyes while you make your escape.

During Krav Maga class, we are trained to move quickly from head control to side control, and then back to head control before possibly switching to side control on the attacker's other side. Always maintain good balance by keeping your legs apart. The laws of physics don't take a rest while you switch techniques. Never separate completely from your attacker when switching between head control and side control, as that would cede control back to your attacker.

360 Outside Defense

I've studied karate for many years, and I love it. Working hard to perform katas well has been one of my favorite parts of karate. Though there are many aspects to katas that will help one in self-defense, katas don't give me the realistic feel of actual fighting that Krav Maga does. Attackers are aggressive, possibly under the influence of drugs or alcohol, and they may be competent-to-highly-skilled fighters. We are taught in Krav Maga class to get in close, inflict damage as fast as possible, and then get free of the danger zone. The "360 outside defense" is one means of getting in close, as I'll describe shortly.

Isaac Newton taught us that a force is required to change an object's state of motion, meaning an object's velocity. Recall that I told you in Chapter 3 that velocity has both magnitude and direction. It's what we in physics call a "vector." Imagine a gruesome scene in which you have to defend yourself against

Figure 4.2
I use a training knife on Mr. Abercrombie. I use an overhead stab in the left photo and an underhand stab in the right photo.

an attacker trying to stab you with a knife. In Figure 4.2, I attempt to stab Mr. Abercrombie with a training knife. I go for an overhead stab in the left photo and an underhand stab in the right photo. In both photos, my knife point shows the vector direction of my knife's velocity. Its velocity changes direction because the knife roughly follows the arc of a circle while I move my arm in a stabbing motion. The magnitude of my knife's velocity is just my knife's speed, which doesn't have a direction. To change my knife's velocity, my knife must experience a net force. My hand is certainly exerting a force on my knife, as is the air my knife passes through. If my knife runs into a fly on the way to Mr. Abercrombie, the fly will exert a force on my knife. Even the sun, moon, and planets exert gravitational forces on my knife, but they are so small that we'll not worry about them. While swinging along an arc of a circle, my knife *has* to feel a net force because my knife's direction changes while moving along that arc. As long as velocity *changes*, meaning speed changes, direction changes, or both change, my knife has to experience a net force.

So what should you do if someone comes at you with a knife? You clearly don't want to get stabbed. It could be that you picked up the attack late and don't have time to sidestep or otherwise evade the knife. The only hope you have of taking control of the situation is to block the knife and counterattack. Krav Maga has a defensive tactic in place for just such an attack—it's called the "360 outside defense." We are taught that such a defense is a "static block against any peripheral attack using the blade of the forearm."[5] Why use the blade of the forearm? If you are involved in a knife attack, the chances of you getting cut are incredibly high. Even the most experienced Krav Maga instructors acknowledge as much. If you are cut on your arm, your chances of walking away from the confrontation with only a minor injury are greatly enhanced if that cut is near bone and not near major arteries and veins. Push on the blade of your forearm and you'll feel your ulna. There is not much to damage between your skin and your ulna. If you get cut on the other side of your forearm, however, you risk having your radial and ulnar arteries cut. Each of those arteries runs along the corresponding bone. Soft tissue and blood vessel cuts are significantly more dangerous than cuts on bone for the simple reason that bleeding out is bad.

The object of the 360 outside defense is to put the blade of your forearm in the path of something dangerous. Blocking means exerting a force, and because you are not deflecting, the force could be large if you are able to stop the motion of the attacker's arm. But you'd have so much adrenaline flowing in an actual attack that you wouldn't notice the large force on your forearm. You would notice that you blocked a knife from cutting you. The "360" part of the defense comes from the idea that you are trying to block attacks coming at you from any direction. Mr. Abercrombie demonstrates the 360 outside defense in Figure 4.3. There are 14 blocks in total, seven on each side. We first practice in the air to get the motion mastered

---

5. Warrior Krav Level 1 Certification Manual.

Figure 4.3
The 14 blocks in the 360 outside defense. Begin in the upper left and move clockwise around until the last photo in the lower left.

before a partner teams up with us and attacks us from random directions. That latter effort is called an "awareness drill," which is "used to develop reflexes and adaptation."[6] If we practice enough, we should be able to reflexively block something we see coming at us. Note that the final two blocks in Figure 4.3 are executed at the same time that Mr. Abercrombie jumps

6. Warrior Krav Level 1 Certification Manual.

back slightly by pushing his backside backward. That moves his body away from a knife stab like the underhanded stab I used on him in the right photo in Figure 4.2.

A defensive block is never used alone. At the same time that you block your opponent's knife arm, you need to violently counterattack. This should be done simultaneously with the block so that not only have you likely avoided a potentially fatal stab, you have possibly turned the tables on your assailant.

Beware of Stick Kinetic Energy

Getting hit with a punch is one thing, but getting hit with a stick, or a tire iron, or a golf club is an entirely different matter. Even a well-executed punch is rarely fatal, but getting hit with a fast-moving, hard, extended object could easily lead to a broken bone or, if you get smashed on the skull, a trip to the morgue. As our Krav Maga manual puts it, "The severity of injuries inflicted as a result of blunt force trauma is dependent on the location and the amount of kinetic energy transferred."[7]

So what is so worrisome about "kinetic energy transferred?" Imagine swinging a stick. Which part of the stick is moving faster—a part near your hand or a part far from your hand? Assuming the stick doesn't break in the air and that rotations happen at your shoulder, elbow, and wrist, the point on the stick farthest from your hand is moving with the largest speed. It has to! Figure 4.4 shows a photo of me swinging my practice stick through the air. I've marked a point at the end of the stick and another point on the stick near my hand. Each circle represents a point separated in time from the next circle by 1/60 of a second. The photo thus shows 4/60 s or roughly 0.07 s of swing time. Both points on the stick moved for the same amount of time, but the point at the end of the stick moved over a greater distance. That means the end of the stick is moving faster than the part of the stick near my hand. Given that kinetic energy scales with the square of speed, there is much more kinetic energy contained in the end of the stick than in the part of the stick near my hand. The speed of each part of the stick is proportional to the radius of the arc it sweeps while in motion. A glance at Figure 4.4 suggests that the outer arc's radius is about twice that of the inner arc. Double the speed and you *quadruple* the kinetic energy. If a fast-moving stick makes contact with a person's limb, some of the stick's kinetic energy

7. Warrior Advanced Krav Maga Phase B Certification Manual.

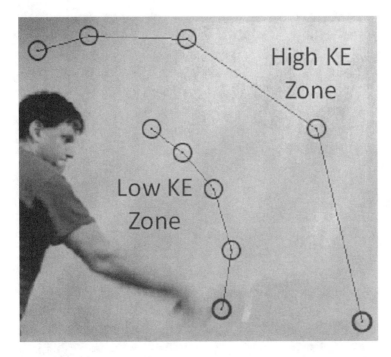

High KE
Zone

Low KE
Zone

Figure 4.4
I swing a training stick. Far
from my hand is the high
kinetic energy (KE) zone;
close to my hand is the low
kinetic energy (KE) zone.

may transfer into "abrasions, contusions, lacerations, and frac-
tures."[8] It takes work to break a bone, and a fast-moving stick
may have enough kinetic energy to transfer into the work
needed to put someone in a cast. Now do you see why "kinetic
energy transferred" is so worrisome?

How do you take control if someone is attacking you with a
stick? You may already know the answer, based on what I just
discussed. You need to avoid the business end of the stick. In
physics terms, the business end is the part of the stick with the
most kinetic energy. Avoiding the stick altogether with a duck
or a step back is one option, but you'd better be prepared for
what comes next once your attacker has swung and missed.
Your attacker won't be satisfied with one swing. You'll need to
quickly and violently counterattack before having to face another swing of the stick. Or you'll need to run, and hope that
your speed and endurance are superior to your attacker's.

8. Warrior Advanced Krav Maga Phase B Certification Manual.

Figure 4.5
I attack Mr. Abercrombie while he demonstrates stick defense.

What will really turn the tables on an assailant with a stick is to rock that person's balance. Instead of ducking or fleeing, close the distance. Your attacker may be surprised to swing a stick at you, only to find you next to his or her head, unleashing a few combatives. Move away from the large kinetic energy zone and into the small kinetic energy zone. Getting hit by a part of the stick or the attacker's arm that was moving along a small radius arc won't hurt nearly as much as taking a blow from the end of the stick. Figure 4.5 shows me about to swing my training stick at Mr. Abercrombie. Note that the training sticks we use are padded to ensure a safe training environment.

Try to put yourself in the mind of someone attacking with a stick. You think, perhaps, that you can strike your victim once or twice, which will allow you a chance to rob the person. Your potential victim is unarmed, and you figure that a good strike to the head will either knock the person out or stun the person long enough for you to make off with the stolen goods. Now

Figure 4.6
Is my victim coming *toward* me?

look at Figure 4.6. Just seeing your potential victim move toward you might rattle you enough to disrupt your balance. The reason I'm getting you to think about being the attacker is that is exactly what I was doing with Mr. Abercrombie. I knew we were in a safe environment and I was using a training stick. But I can assure you that seeing someone come toward you when you attempt to strike that person can be a tad off-putting the first time or two it happens. Part of training is to gain the perspective of the attacker. As a physicist, I love attacking because I can see firsthand how physics works against my attack. What I can see in Figure 4.6, and what I was processing while attacking Mr. Abercrombie, is how balanced he is in his stance. If he plans to disrupt my balance, his balance better be good. His shoulders are square, his arms are up and moving toward me, and his eyes are locked on me. His technique is nearly flawless, which is why I was lucky to be in a training session.

It's now time for Mr. Abercrombie to burst into the low ki-

Figure 4.7
Mr. Abercrombie has closed the distance and tables are about to be turned.

netic energy zone. Figure 4.7 shows that I haven't even gotten my stick rotated toward my potential victim. Mr. Abercrombie bursts in with his left arm shielding his body from the fast-moving stick and his right hand delivering an eye poke (just simulated in training). My right arm and stick haven't yet made contact with Mr. Abercrombie.

When I make contact with Mr. Abercrombie in Figure 4.8, I have hit him too close to my shoulder and elbow. In other words, the arc radii of the parts of the stick and my arm that made contact with Mr. Abercrombie were too short to do any damage. Look at how far behind him the business end of my stick is! There is now no hope of me transferring my stick's kinetic energy to my potential victim. A lot of that kinetic energy got transferred into my stretched tendons and ligaments as the stick came to rest.

Do you recognize side control in Figure 4.9? At this point, Mr. Abercrombie has immobilized my weapon limb with an

Figure 4.8
Mr. Abercrombie is in the low kinetic energy zone and I'm already regretting attacking him.

Figure 4.9
Mr. Abercrombie has me in side control, and convinced me that I took a swing at the wrong person.

Figure 4.10
I'm about to take a swing at Mr. Abercrombie with a backhand stick technique.

overhook. What comes next is an aggressive counterattack, probably consisting of at least five combatives. He's in a great position to deliver a knee strike to my abdomen, a leg strike to my groin, an elbow to my face, and other such niceties. Think how physics helps explain the effectiveness of the stick defense. By quickly closing the distance between himself and me, Mr. Abercrombie moved out of the high kinetic energy zone and into a region where I couldn't possibly transfer much energy to his body. Knowing where to move to avoid that kinetic energy transfer further allowed Mr. Abercrombie to secure me in side control. His combatives might even dislodge the stick from my hand, and he may be the one with the weapon when I finally get my bearings after being pummeled. A firm grounding in the basic physics principles that explain why a given technique is so effective will, in my humble opinion, help you perform the technique better. When Mr. Abercrombie first showed me a stick defense, I immediately thought, "Oh I see, I need to get where the stick's kinetic energy isn't so big." I still

Figure 4.11
Mr. Abercrombie steps into
the low kinetic energy zone.

need to practice the technique many, many times to feel confident enough that I could employ it if I ever had to. Understanding the physics helps me visualize what I need to do and makes
my training more efficient. At least that's my hope.

Now you're an expert on the difference between a low kinetic energy zone and a high kinetic energy zone. Suppose
someone attacks you with a stick using a backhand technique,
as in Figure 4.10. You got surprised and the stick is coming too
fast for you to safely duck out of the way. What do you do?
Think about it before reading any further.

If you said, "Move forward into the low kinetic energy zone,"
give yourself a well-deserved pat on the back. Look at Figure
4.11. Mr. Abercrombie is already moving forward when I've
only cocked my stick behind me. This is the key to taking back
control from your attacker. Don't let somebody think he or she
can just strike you with a stick. Be proactive and let your attacker know that you won't be messed with. Turn the tables
quickly before you sustain an injury. Note in Figure 4.11 that

Figure 4.12
Having stepped into the low kinetic energy zone, Mr. Abercrombie has turned my dangerous backhanded stick into a painful forearm for me.

Figure 4.13
After rotating into me, Mr. Abercrombie has made me aware that once again, I've messed with the wrong person.

Mr. Abercrombie is already preparing to be hit on his right side. His left hand is protecting his chest as he steps forward. And most important, his eyes are on his attacker. Techniques are far more effective if you watch what you're doing.

Figure 4.12 shows exactly what you were expecting. What I thought would be a debilitating backhand with a stick turned into a bruised forearm and a harmless stick behind my intended victim. Mr. Abercrombie's right arm is already moving up for what comes next. In Figure 4.13, Mr. Abercrombie rotates clockwise, meaning he rotates into me. He uses his right arm to immobilize my limb that holds the stick, and his left arm shoots across my face. This is how you take control back from your attacker.

From holding me in a controlled manner, Mr. Abercrombie now rotates back in the direction from which he came. His left arm is poised to rotate my stick-wielding arm away from him, and move me into an all-too-familiar position. Check out Figure 4.14 as I'm being rotated to a position where I definitely

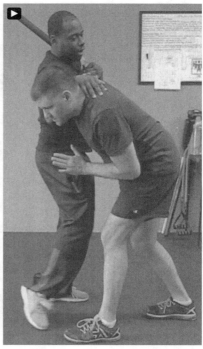

won't be in control. This technique is especially useful if your attacker continues to fight you as you hold him or her, as in Figure 4.13. If your attacker continues to be aggressive toward you, getting that person into side control and unleashing a few combatives will take that aggression away in a hurry.

The confrontation ends in Figure 4.15 as Mr. Abercrombie has put me in side control. From the looks of the photo, I'm about to receive a knee strike to my solar plexus. All techniques are practiced in a safe environment. But I hope you can see how to take control back from an attacker with a stick. It doesn't matter whether the attack comes via forehand or backhand. Physics tells us to move in toward the low kinetic energy zone and go from there.

I'll end this section with another physics question. Will the ideas behind the low and high kinetic energy zones of a swinging stick be limited to attackers with sticks? Of course not. Kinetic energy involves the product of mass with speed squared, and it matters not what the object is, as long as it has mass and

Figure 4.14
If I continue to fight Mr. Abercrombie, he can rotate back toward his original position, keeping my stick limb in control as he rotates me.

Figure 4.15
Side control means that as the attacker I have lost control to my potential victim.

Figure 4.16
My swing at Mr. Abercrombie turned out to be a poor decision on my part when he stepped into my punch while protecting his head with reflexive cover.

it's in motion. Such an object could be an arm swung with a fist at the end. Suppose I were to take a big roundhouse swing at Mr. Abercrombie. He'll make use of what we call "reflexive cover," which is a "structural defense used to absorb blows to the head and body."[9] In Figure 4.16, you see that Mr. Abercrombie has put his right hand behind his head and has his right arm tight against his face. My sweeping left roundhouse punch was intercepted by his arm at my upper arm. That is the low kinetic energy zone for my arm. The business end of my arm, my fist, is harmlessly behind Mr. Abercrombie's head. What comes next? All he has to do is unwrap his right arm, hook over my left arm, shoot his left arm across the left side of my face, perhaps getting my eye along the way, and he'll have me in side control. He could also bring his right arm under my left arm and achieve side control that way. I don't need to show you a photo of what comes next. By now, you have side control in your mind and you can visualize it.

9. Warrior Krav Level 1 Certification Manual.

Making Use of Leverage

What is the first thing you think of upon hearing or reading the word "leverage?" Do you think of borrowed
money and all its economic implications? Or do you think
about the effective use of power and influence in gaining advantage in a given situation? For me, I think about the first five
letters of "leverage," *lever,* which is one of the so-called six simple machines[1] identified by top thinkers in the Renaissance
that provide one with what is called "mechanical advantage."
That term refers to the ability of a machine, or person in the
case of our discussions in this book, to multiply force. Leverage
is what allows a smaller, weaker person to defeat a larger, stronger attacker. Understanding the physics behind mechanical advantage will help you understand how various Krav Maga techniques are so effective.

Mechanical Advantage Basics

You have made use of "mechanical advantage" your entire life,
even if you've never thought about or heard the term until now.
Ever been on a teeter-totter? Used a wrench or crowbar? You've
opened a door, right? From tools to toys, mechanical advantage
is everywhere. I introduced torque in Chapter 2 as the product
of a lever-arm distance and a force. A torque is required to
change an object's rotational velocity, just as a force is required
to change an object's linear velocity. Changing rotational velocity means rotational acceleration is present, just as changing

1. The other five are the inclined plane, the pulley, the screw, the wedge, and the
wheel and axle.

linear velocity means linear acceleration is present. One of the many beauties of physics is that, although we can't argue with nature about how much torque is required for a given angular acceleration, we have two parameters to play with to create that required torque. If we find we can't exert enough force to generate the necessary torque, we can change lever-arm distance. If you've ever struggled to loosen a bolt with your fingers, you know where I'm headed.

Check out Figure 5.1. The left photo shows a close-up of my fingers trying to loosen a 1.5-in diameter bolt. Because the rotation axis passes right through the center of the bolt (and perpendicular to the page), the lever-arm distance is 0.75 in, or the radius of the hexagonal bolt head.[2] Such a small radius requires a large force to get the torque needed to loosen the bolt.

The solution to the problem of loosening a bolt is shown in the right photo of Figure 5.1. The wrench provides a much longer lever-arm distance (15 inches, in the case of my wrench) than using only my fingers. For the same force applied to the bolt head, the mechanical advantage provided by my wrench is simply (15 in / 0.75 in) = 20. The wrench thus allows me to exert just 1/20 or 5% of the force I would need using only my fingers. Anyone who has made use of a wrench already knows what an enormous advantage it is to increase the lever-arm distance. There is no free lunch, however, because if it took one complete turn to loosen the bolt, my hand on the wrench's end would have to travel 20 times farther than my fingers would need to travel if they could loosen the bolt. The circumference of the circle my hand would trace out while using the wrench is

2. Because my fingers apply force at the points, or vertices, of the hexagon, I measured the maximum diameter. The minimum diameter is the distance between opposite faces of the hexagon. It's a teeny bit of geometry to show that the maximum diameter is $2/\sqrt{3} \simeq 1.155$ times the minimum diameter. The minimum diameter is what's specified on the wrench I used. Note that it's much easier to measure diameter than it is to measure radius. Lining the edge of a ruler or a caliper on the edge of the bolt head is not as challenging as trying to eyeball where the center of the bolt is.

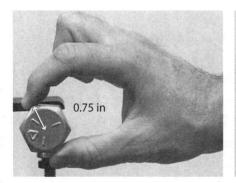
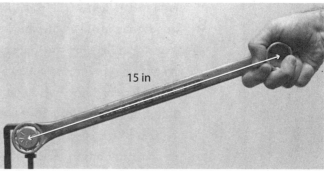

roughly 47 in or nearly 4 ft. If my fingers could loosen the bolt without the aid of the wrench, they would only need to move through a circle of circumference of nearly 2.4 in. But none of us is likely to think of moving through 20 times as much distance as a problem, especially if we can't muster enough strength to loosen the bolt without using the wrench. In the same vein, I doubt you'll complain if you have to perform a Krav Maga technique on a big attacker if reducing the force necessary to subdue your attacker means you have to move through a greater distance.

Figure 5.1
A wrench provides a large mechanical advantage.

It Doesn't Take Much Force

Have you ever had your arm or leg caught in something that causes you to feel very uncomfortable? Even a seemingly silly situation as getting your arm caught in a jacket can lead to frustration and possibly a tiny bit of pain as you try to rescue your arm. Staying as far from the technical jargon of an anatomist as possible, there are ways in which parts of our bodies aren't meant to move. There was an evolutionary advantage to develop the joints we have. When our ancestors evolved from water based to land based, we needed ways to move around on the ground quickly. If we had to drag our bodies, friction would slow us down too much and we couldn't outrun predators. Limb joints allowed us to vary how much our bodies were in contact with the ground. The ball-and-socket joints that make up our shoulders and hips allow for great range of motion. The hinge joints in our elbows and knees are less flexible, but they're great for collapsing the lengths of our arms and legs, respectively. But hyperextend your arm or leg, and you could be heading for lots of pain. Twist your arm or leg with a rotation axis along your arm or leg and you could be screaming in pain.

Krav Maga, even if it isn't consciously doing so, exploits the parts of our bodies where we haven't evolved with much range of motion. The technique I'll describe next involves making use of leverage by rotating an attacker's arm in a way that the arm isn't meant to rotate.

What if someone grabs your shirt or lapel? That's a fairly modest assault when compared with other attacks I discuss in this book, but it's still unwelcome contact from an aggressive person. An attacker's intent is always something you should worry about. Maybe someone didn't like the way you looked at him or her; maybe you accidentally bumped into someone. Regardless of any perceived slight on the part of the belligerent stranger who grabbed your shirt or lapel, there is no legitimate reason for someone to grab your clothing. Suppose you were grabbed so fast that you didn't have the opportunity to intercept the stranger's hand. If the person doesn't appear overly hostile or violent, you can say, "Please remove your hand from me." If that doesn't work, you may be in a situation in which a physical confrontation is unavoidable. You'll be in that situation for sure if, before you can make a polite request, you are pushed or pulled. You need to react quickly before your assailant takes the confrontation to a much more violent level. The physics of leverage will be on your side.

The situation I just described is depicted in Figure 5.2. I grab my instructor's, Mr. Abercrombie's, shirt. We practice scenarios in which the attacker is pushing and pulling with much more aggression than what I show in the photo. But the technique is the same nevertheless. What do you think Mr. Abercrombie should do? You've been introduced to the Renaissance idea of

Figure 5.2
I grab Mr. Abercrombie's
shirt.

simple machines, to a basic idea of leverage, and to the notion that Krav Maga techniques might exploit twists and turns that our bodies haven't evolved to welcome. Should he push my arm aside? Despite the fact that Mr. Abercrombie is much stronger than I am, if he tried pushing my hand, he would be using a short lever arm to exert a lot of force for the torque needed to swing my arm off his shirt. I could bend my arm slightly and rotate my torso in such a way that even my strong instructor would be struggling to push my arm away. Pushing my arm away by applying force to my hand also means trying to rotate an object, my arm, that has a lot of rotational inertia, meaning that much of my arm's mass is far from where Mr. Abercrombie would be pushing. The key is to find a low-force option that will twist my arm in a way it's not supposed to rotate.

Our fingers give us great gripping ability, especially when used together and in combination with the palm. Our ances-

Figure 5.3
Mr. Abercrombie turns my hand in a direction that will make it tough for me to continue holding his shirt.

Figure 5.4
My hand is off Mr. Abercrombie's shirt and he has control of my wrist.

tors needed that when they spent time in trees. But we don't have as much strength *between* our fingers, which means we can't pinch with much force. Sure, it hurts if someone pinches you with a thumb and forefinger, but it's easy to pull away. Imagine someone holding a stick. It's much easier to pull the stick out of the person's hand if you pull in a direction away from the crotch of the thumb and forefinger than if you pull on the stick as if you were playing tug-of-war with the person. That realization offers an insight into my grip on Mr. Abercrombie's shirt. If he tried pulling away, we'd be playing tug-of-war with his shirt. He might win, but I could at least put up a good fight.

Mr. Abercrombie's self-defense begins by turning my hand in a direction that makes it more difficult for me to hold onto his shirt. Check out Figure 5.3. Mr. Abercrombie has reached over my arm with his right hand and grabbed my wrist. Instead of

Figure 5.5
Turning my arm in a
direction it wasn't meant to
turn is like turning a hex key.

rotating my arm such that a rotation axis is vertical and through
my shoulder, Mr. Abercrombie plans to rotate my arm with a
rotation axis right through my arm. There is much less moment
of inertia to overcome with such a rotation because the mass in
my arm isn't very far from the rotation axis.

You can see in Figure 5.3 that Mr. Abercrombie's left hand
is moving upward. If he doesn't secure my hand, he risks me
pulling my arm away by exploiting the same pinching weakness
that he's using to get my hand off his shirt. Figure 5.4 shows
how Mr. Abercrombie has control of my wrist using both of his
hands. There is no direction where I can easily pull my arm free.
By plucking my hand from the left side of his body with his
right hand, Mr. Abercrombie's left hand not only secures my
wrist, it provides a natural rotation direction, which is clockwise (as seen by him). Mr. Abercrombie also makes use of lever-

age by grabbing my wrist and using it to rotate my arm. He's using a longer lever arm than if he had tried turning my arm by twisting my forearm. That increase in lever-arm distance reduces the amount of force he needs to twist my arm.

Mr. Abercrombie simply keeps twisting my arm clockwise (from his viewpoint). His thumbs are securely clamped on the back of my hand. By turning my hand beyond my wrist joint, Mr. Abercrombie exploits a larger lever-arm distance than if he turned my wrist. The cross section of my wrist isn't a circle, but I estimate a mechanical advantage in the range of 2.5–3.0 for turning my arm using my wrist-metacarpophalangeal (big knuckle) distance compared to turning my arm using just my wrist. Turning with the hand instead of the wrist confers advantages over and above the mechanical advantage numbers I gave. It's far easier to hold onto the hand than the wrist, partly because an attacker's hand is no longer free to fight you. For the mechanical advantage, imagine turning a hex key with a 90° bend in it; this bend greatly increases lever-arm distance and makes turning hexagonal-socketed screws a breeze. This technique is so much like turning a hex key that I've added an image of a hex key to Figure 5.5. My arm is bent just like that hex key. That's why it's crucial to bend the attacker's hand. You need that added lever-arm distance so that you can easily execute the technique without the need for lots of strength. The technique wouldn't be nearly as effective if an attacker's arm were comfortably bent at the elbow. This is how leverage works to your advantage, especially if your attacker is much bigger and stronger than you. Physics has a solution to the problem of being bullied by a big thug.

Note in Figure 5.5 that Mr. Abercrombie has stepped back with his right leg. That serves two purposes. By moving his center of mass backward, and carrying me in the process, I'm forced to lean over more. I'm already suffering from having my arm twisted in a way that evolution didn't prepare me for.

Any lateral movement is going to cause me to move with my former victim because I don't want my arm hurting any more than it already is. The second purpose in stepping back is that Mr. Abercrombie is in prime position to deliver a kick with his right leg if I haven't learned my lesson yet. Because he has control of my right arm, Mr. Abercrombie is more stable if he kicks with his right leg. All he has to do is pull my constrained hand down and in toward his abdomen at the same time he sends a front kick to my face or throat. If I get kicked in the face and I still want to mess with Mr. Abercrombie, I will be learning physics lessons the hard way.

Leverage for Restraint Instead of Pain

Leverage may be used to restrain as well as cause pain. Suppose I was intoxicated when I grabbed Mr. Abercrombie. He could smell the booze on my breath, and he guessed that I might not be so aggressive were I not drunk. I'm certainly a danger to him, others, and likely myself, but I haven't pulled a knife or a gun. So kicking me in the face or throat, as might have been the end of the confrontation I described in the previous section, could be viewed by some as over-the-top. I'm certainly in the wrong if I'm drunk and messing with people, but if I'm not acting like a crazed madman, bashing in my face while my reflexes are dulled and my wits aren't with me is excessive force. A better option might be to find some way to restrain me until proper authorities arrive and arrest me for drunk and disorderly conduct.

Start back at Figure 5.2, in which I'm grabbing Mr. Abercrombie's shirt. His goal is no longer to get me in a position to administer punishment, it's to get me in a position where he can restrain me. Instead of reaching across with his right hand, Mr. Abercrombie uses his left arm to simultaneously remove my hand from his shirt and get his left arm ready to restrain me. He initiated the technique by making a 90° angle with his

Figure 5.6
Restraint, not incapacitation, will be the goal.

Figure 5.7
Will it be side control or restraint?

left elbow and rotating his left arm clockwise (as seen by him) on top of my shirt-holding right arm. Figure 5.6 shows Mr. Abercrombie's left arm rotating over my forearm.

As Mr. Abercrombie rotates his left arm, he steps and leans toward me as his right arm comes up. Figure 5.7 shows that I could easily be put into side control at this point, but Mr. Abercrombie has restraint on his mind. His eyes are locked on my right shoulder. He wants to get behind me.

Now we get to the action in Figure 5.8. Instead of moving straight into me and putting me in side control, Mr. Abercrombie is rotating around toward the back of my right shoulder. Note that he has maintained control of my right arm. His right arm is up near my neck. That serves to keep me off-balance and to help him rotate. Look at my right elbow. All Mr. Abercrombie has to do next is complete his rotation and my right arm will bend according to my elbow's desired hinge direction.

The completion of the technique is shown in Figure 5.9. Mr. Abercrombie is behind me, having kept his left arm clamped on mine as he rotated back. My right elbow is comfortably bent, but because Mr. Abercrombie has my right arm locked from below, all he has to do is raise up his left arm and I'll become uncomfortable in a hurry. He is making superb use of leverage as the crotch of his left elbow is under my right wrist. If he lifts up, he'll have a long lever arm between my wrist and my shoulder. My arm wasn't meant to go much higher up my back, and I'll want to cooperate or face serious pain. The fact that Mr. Abercrombie has secured my left shoulder with his right hand means that he has even more of a leverage advantage on me. If he were to raise his left arm, he could simultaneously pull my right shoulder backward. His left wrist makes a perfect fulcrum in my back so that my upper torso can rotate backward, which will intensify the pain in my right shoulder.

Figure 5.8
No side control. Mr. Abercrombie heads behind me.

Figure 5.9
I've been restrained, and someone has called the police.

The aforementioned restraining technique involved multiple applications of leverage. Always keep in mind that a torque is the product of a force with a lever-arm distance. By creating effective simple machines like a lever (my arm) or a wheel (his rotating arm), large torques may be generated from small forces. Without the ability to use physics and leverage, a smaller and weaker person would almost always be at a disadvantage when attacked by a bigger and stronger person. The next technique I will discuss really makes that point because, without leverage, doom is the only outcome.

A Bridge to Freedom

One type of situation we have a tendency to panic or freeze is one in which we experience a sense of claustrophobia. Being trapped against a wall can give you a sense of "no way out." Another such scenario is being pinned to the floor. We are taught in Krav Maga class that the intent of an attacker is the real danger in a confrontation. Does your attacker intend to steal from you? Harm you? Sexually assault you? Take you to another location? Never assume that, for example, pinning you to the floor is the extent of what your attacker has in mind. You must overcome the sense of panic you feel in such a harrowing and claustrophobic situation, and react in a way that disrupts your attacker's balance.

An example of being pinned to the floor is shown in Figure 5.10. In the left photo, Ms. Maupin tries in vain to remove Mr. Abercrombie from on top of her. The problem is that she is trying to get him off of her by using her arms. She is trying to move too much mass from too great a distance. A long lever arm—her arms' length, in this case—means a lot of torque is needed to rotate Mr. Abercrombie off her. The right photo shows that if she first gets into a bridge position[3] with her hips,

3. To get into a bridge position, lie down on your back and lift your hips and pelvic region off the ground so that your shoulders support a significant fraction of your body weight.

she will be in a position to use leverage against Mr. Abercrombie. Back, hip, and leg muscles are better to use in this situation than arm muscles.

Figure 5.10
Arms, no. Bridge position, yes.

Leverage is a key component to fending off an attacker if you are significantly smaller than your assailant. Neither I nor Ms. Maupin will have much success if we try to get into a test of strength against Mr. Abercrombie. Instead of wrestling with your arms, take advantage of the strong muscles in your hips, and move a strong attacker a short distance so that leverage can set you free.

I'll let Mr. Abercrombie demonstrate the technique of freeing yourself from an attacker who is in a mounted position from above. Figure 5.11 shows me on top of Mr. Abercrombie. I have my hands to his throat to further intensify the seriousness of the attack.

Mr. Abercrombie is certainly much stronger than I am. But even he doesn't want to arm wrestle me from the position he's in. I don't have his strength, but I do have the advantage of being able to swing my arms and create momentum and kinetic energy in my fists before hitting him. The floor prevents him from cocking his arms back or generating much momentum and kinetic energy in a punch.

Notice what has put Mr. Abercrombie in immediate danger—my hands are at his throat. Even if my intent is something much more pernicious, he needs to get my hands off his throat. The beauty of Mr. Abercrombie's first move is that it both aids his breathing and sets up his subsequent moves. Check out Figure 5.12. By throwing his left arm across my arms, Mr. Abercrombie is able to create just enough space at his throat to allow him to breathe. His right hand grabs my left triceps area. These moves offer two advantages: his right hand didn't have to travel very far, which minimizes technique time; and he now has my arms pinned in such a way that it will be difficult for me to continue choking him. He can push down with his left forearm, which will cause my elbows to buckle. That's simple physics in action. My elbows provide natural fulcra. I have to fight against his left forearm, and now my range of motion is constricted by having my left upper arm pinned.

Now comes a very important step in the technique. Whichever arm the defender used to grab the attacker's triceps is the side of the defender's body toward which the attacker will be rolled. Figure 5.13 shows how to further enhance leverage, as Mr. Abercrombie uses his right leg to hook my left foot. Note the subtle difference between Figure 5.12 and Figure 5.13. By moving my left foot toward the center of our bodies, I've lost stability on my left side. I no longer have my left knee and my left foot outside of the majority of my mass. To rotate me, all Mr. Abercrombie has to do is use my left knee as a fulcrum.

Think more carefully about what just happened after my left foot got hooked. I'll use an analogy. Is it easier to push a refrigerator over while it's standing up or while it's on its side? Though you may need some decent strength to do it, it's much easier to turn the upright refrigerator over. Its base isn't as wide as it would be if it were lying on its side. As I discussed in Chapter 2, stability is lost if an object's center of mass is pushed past a point of contact with the ground. Now suppose instead of pushing an upright refrigerator over while it sits on a flat floor, the floor could sink on one side of the refrigerator. If you push toward that side, gravity will help you topple the refrigerator even more easily.

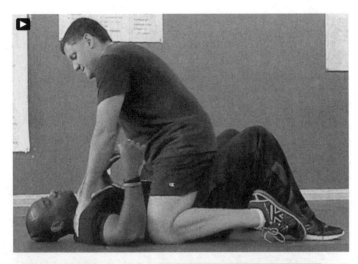

Figure 5.11
I have the advantage over
Mr. Abercrombie—or do I?

Figure 5.12
Mr. Abercrombie begins
his defense against a
choke from an attacker in a
mounted position.

Figure 5.13
Mr. Abercrombie hooks my
left foot.

Figure 5.14
Bridging makes it easier to topple me.

Figure 5.15
Mr. Abercrombie rotates to his right, thus turning the tables on his attacker.

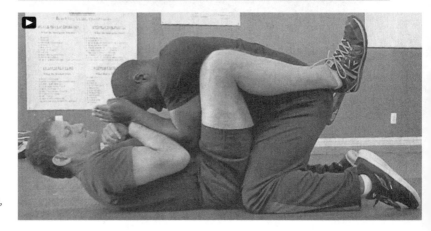

Figure 5.16
Tables have been turned, but Mr. Abercrombie still protects his face. Though he may have won the battle, the war may rage on.

With the previous idea in your mind, look at Figure 5.14. Mr. Abercrombie has bridged upward. His hips have pushed me up. You can see that my left knee is no longer in contact with the ground. Had I been much taller and Mr. Abercrombie had been much smaller, my knee might have still been in contact with the ground. Ms. Maupin was in that situation in Figure 5.10. Either way, having my left leg locked closer to our centers makes for a smaller moment of inertia, which makes it easier to topple me. Even a very small person can do this. All that's needed is just a little upward adjustment to make toppling your attacker easier. Mr. Abercrombie needs only to drop the right side of his hip and let gravity help move my center of mass to the point that I lose stability. It is, after all, easier to topple an object on an incline (or a refrigerator toward a sunken floor) than it is to topple an object on a flat surface. Mr. Abercrombie has now gained a mechanical advantage. His force and torque will be multiplied because of the help provided by gravity.

Now we get to the point where the tables are turned on the attacker. In Figure 5.15, having secured my arms, Mr. Abercrombie now drops his right hip and rotates toward his right. Note that my left foot is still locked. You can see that Mr. Abercrombie didn't have to drop his right hip very much and he didn't have to turn very far toward his right to get me into an unstable position. My balance is being disrupted, and I'm about to fall to the mat.

Figure 5.16 shows the completion of the technique. Having turned me over, I am now the one on the ground, possibly feeling a little claustrophobic myself. Once he's on top of me, Mr. Abercrombie has immediately put his arms up to guard his face. After all, I may not be completely disoriented, and I might react by punching him. Even worse, I may have been able to get a knife from my waistband with my right hand. Mr. Abercrombie has his ulnas facing outward in case his arms get cut. As we discussed, it's better to get cut on bone than on fleshy parts with life-sustaining arteries and veins.

A series of techniques could follow what you see in Figure 5.16. If you are ever unfortunate enough to find yourself pinned to the ground, and you manage to successfully execute what I've just described, you'll be lucky if you can just hop up from the position in Figure 5.16 and run away. Sure, it's possible that your attacker is so drunk or disoriented that you have a window of time to escape. If the confrontation continues, however, you will need to administer combatives immediately after turning the tables. You should at least offer a knee to the groin and/or a stomp to the groin upon getting up. One combative is better than none. If you're unlucky enough to have an experienced fighter attacking you, more combatives and more sophisticated techniques may be required. But what I've described here allows you to use good physics-based techniques to get out of being pinned to the ground with an attacker's hands at your throat.

From Lug Wrench to Clavicle

Suppose you are engaged with an attacker, and you manage to turn the tables and get your assailant in side control. Maybe your attacker is drunk, and you conclude that driving a hammer fist into the spine or a knee into the face might be a bit much. One alternative is a simple takedown. This allows you to restrain, administer more punishment, or just flee the scene as fast as possible. In fact, takedowns are useful against good fighters who aren't drunk. You may find a very brief opening that allows you to get your opponent on the ground. That may be the best option you have in defeating a skilled opponent. We are taught to avoid getting into a ground fight because there are so many added risks when the fight gets taken to the ground, such as, "Asphalt is hard, it scrapes the skin, and is unforgiving" and "You can receive double impact from a strike—the strike

itself, and the floor."[4] We think of getting someone on the ground as putting ourselves in a dominant position. But if we are also on the ground, we need to focus on survival instead of fighting.

We learn Takedown #1 after we've progressed a couple of belts in our Krav Maga training. Performing takedowns is not recommended for beginning students because they need to become acclimated to Krav Maga before taking an opponent (or training partner) to the mat. Safety comes first! When I was finally introduced to Takedown #1, the first image that popped in my mind was a lug wrench. Once you have someone in side control, as in Figure 4.1, it's a simple transition to getting your opponent to the ground.

Figure 5.17
Mr. Abercrombie is about to take me down in the left photo. I'm using a spider-type lug wrench in the right photo. The arrows show force directions.

4. Warrior Krav Advanced Phase B Certification Manual.

Check out Figure 5.17. Mr. Abercrombie had me in side control and then began executing Takedown #1 in the left photo. My right clavicle and the upper part of my right humerus make up the majority of the distance between the two points on my body where Mr. Abercrombie applied force. His left arm pushed my right arm upward and his right hand pushed my neck downward. The force he applied with his right hand was slightly larger than that from his left arm. Suppose they were the same. He may not have been exerting a net force on me given that his upward force would match his downward force, but because the two forces weren't along the same line, he was exerting a net *torque* on me.

Now look at the right photo in Figure 5.17. I'm tightening one of the bolts in my back tire with a spider-type lug wrench. I pushed up with my left hand and down with my right hand. Ignoring the small net upward force on the wrench I had to apply to keep it from falling, the two forces I applied were essentially the same. Why did I need a lug wrench? Because the added lever-arm distance gave me a mechanical advantage of about 12. I generated a large torque on the bolt and got it nice and tight. Mr. Abercrombie took advantage of the added lever-arm distance to exert a large enough torque on me to get me to rotate clockwise (as seen by him). Once I started rotating, my balance was lost and the gravitational tug from Earth helped Mr. Abercrombie get me to the mat. Unlike the technique I discussed in the previous section, I don't need to show you several photos of the move. The photo in Figure 5.17 has it all. Once you have an attacker in side control and you want to get that person to the ground, just pull up with the arm you have wrapped on your attacker's arm and push down on the attacker's neck. The leverage provided by the lever-arm distance allows you to generate a large torque.

Linear Momentum and the Flyby

Yet another physics word pervades our vernacular. That word is "momentum." You've heard sportscasters say something like, "After winning six in a row, the Yankees have all the momentum entering the weekend series with the hapless Tigers." Or perhaps, "The Spurs crush the Heat in Game 3! The momentum has switched after the Heat's narrow Game 2 win." Common usage of the word "momentum" is actually not too far off from the physics meaning. We in physics make the definition of momentum quite clear, but the notion of movement and direction are good starting points for an understanding of momentum. The Yankees are coming to town on a six-game winning streak. The Spurs head into the fourth game of the Finals full of confidence. Teams with massive athletes collectively moving toward a desired goal is a good way to visualize momentum.

Newton's Second Law may be expressed in terms of momentum or, more accurately, *linear momentum*. Newton himself thought of changes in motion in the context of what we now call linear momentum, which is simply the product of an object's mass and its velocity. The latter part, or the velocity, is what gives momentum[1] its direction. Think about what is contained in that definition of momentum. By combining both mass and velocity, we have a quantity that gives us a feeling for getting hit. In other words, would you rather be hit by a ping pong ball moving 5 mph or a bowling ball moving 5 mph? Despite the equivalent velocities, the bowling ball's greater mass

6 CHAPTER

---

1. I'll stick with the word "momentum" in this chapter. If there is possible confusion with "angular momentum," I'll be sure to tack on "linear" in front of "momentum."

makes it scarier. Would you rather be hit by a fist moving 0.1 mph or a fist moving 10 mph? The masses are the same, but we fear the faster moving fist. Momentum nicely encapsulates an object's size and how fast it's moving.

Yet the significance of momentum in physics goes beyond our concerns about being hit by objects. Newton's Second Law, which I described in Chapter 1, equates an object's mass times its acceleration to the net, external force an object feels. In other words, the net, external force that an object feels is the time rate of change of its momentum. That means that if an object feels no net, external force, its momentum can't change. It also means that, to change an object's momentum, a net, external force must be applied to that object. The reason you don't want to get hit by a fist moving at 10 mph, but might not care about being hit by a fist moving at 0.1 mph, is that the faster moving fist takes 100 times more force to stop than the slower moving fist, if the stopping times are the same. If your jaw has to do the stopping, that fast-moving fist is going to hurt.

Collisions are where momentum gets really useful. It turns out that for multiple objects, such as two fighters, if there is no net, external force on the *system* of objects, that system's momentum can't change. This is the *conservation of linear momentum* that is used to help understand car collisions, sumo collisions, and collisions between Krav Maga combatants. Two fighters on a mat certainly have Earth pulling down on them (forces we call their weights) and upward forces from the mat. They also have horizontal friction forces from the mat, but if we are analyzing collisions that last only a short time, we can ignore the small friction forces compared to the forces felt by one fighter on the other.

Have you ever played pool or billiards? If you hit the cue ball just right, you can have it hit another ball, transfer its velocity to the other ball, and leave the cue ball at rest after the collision. That ensures system momentum is the same before and after the collision. If you hit a ball off-center, one ball moves at an angle to the left while the other ball moves at an angle to the right. That also keeps system momentum conserved.

The units for momentum are likely unfamiliar to you. But because units follow the same algebraic rules as numbers, it's easy to form units in various systems. Just multiply a mass unit by a velocity unit. Most of the world uses the SI system of units, which means multiplying a kg by m/s to get linear momentum units. The British Imperial units use the "slug" for mass and ft/s for velocity. I lived in England for two years, and I don't think I ever heard a person refer to mass in slug units. Recall from Chapter 1 that a force unit is a mass unit times an acceleration unit. The force unit in the British Imperial system, the pound, is the same as a slug times a ft/s^2. If you're from the United States, you use pounds all the time, but you probably didn't know that there is a slug lurking in that pound! A slug is a decent-sized unit that's equivalent to about 14.6 kg and weighs approximately 32.2 pounds. Think of a slug as two maximum-weight bowling balls. I have a mass of around 6 slugs. Can you find your mass in slugs? Keep in mind that *any* mass unit times *any* velocity unit will give a momentum unit. Though not terribly useful, a slug times a furlong per week is a momentum unit. I want you to be aware of units that are used for momentum, but you really don't need that knowledge for what follows. A great deal of physics may be learned and applied without crunching any numbers and keeping track of units. I will, however, leave you with one quick exercise. You know how force units are made and you now know how momentum units are made. Convince yourself that units of force times units of time give momentum units. You'll need that concept soon.

The aforementioned momentum ideas will help us understand various collisions in Krav Maga. They also help us understand why we want to avoid a collision altogether, as the next section will illustrate.

The Flyby

Imagine someone running at you with a knife. The person's mass times the person's velocity is your attacker's momentum. You won't be thinking about such a concept if you're ever unlucky enough to be in such a situation, but thinking about momentum now could save your life. Should you try to stop your attacker? Recall that changing momentum requires a force, and stopping someone from running at you means slowing the attacker all the way down to zero speed, which could require a significant amount of force, depending on how massive your attacker is and how much time is used to stop your attacker. Don't outsmart yourself. Sometimes the simplest move is the best option. If someone is charging at you, there is no need to ponder all the various takedown techniques you've learned. Just get out of the way. Look at Figure 6.1.

Figure 6.1
Step aside and let linear momentum save your life.

If you step aside, an attacker will need to change the direction of his or her momentum in order to pursue you, and that takes substantial force if the attacker is moving fast. In that case, the net force exerted by the attacker will likely come from friction with the ground—but that's the attacker's problem. Contrast this with the net force that you would have to exert to bring him or her to a complete stop. Don't try to exert that force, and definitely don't get in the way of a knife. Getting out of the way is far better than bracing for a collision.

Other ways of making good use of linear momentum include simple side movements of the head while being punched and various types of parry, or deflect, techniques.

Bend Those Knees

I'm now going to convince you yet again that you have some intuitive physics that has served you well in your life. Have you ever climbed over a fence? Jumped off a big rock? Leaped off your couch when your team makes the play of a lifetime to win the big game? Ever thought about how you landed all those times you jumped off something and touched down feetfirst on the ground? Probably not, but let me give you a thought experiment.[2] Imagine standing on a chair, jumping off, and landing on the ground with your legs straight, meaning your knees aren't bent. Even if you've never studied physics before, you likely know that landing on the ground like that would hurt your knees and perhaps your hips. All those times you climbed over a fence, jumped off a big rock, and leaped from the couch in celebration, you bent your knees. You had good intuitive physics.

You bend your knees upon landing on the ground to extend the collision time with the ground. Just before landing, you

---

2. Ponder this, but don't try it!

have a certain momentum, which is the product of your mass with your velocity. When you hit the ground you come to a stop, which means your velocity is zero, hence your momentum is zero. It turns out that your change in momentum is equal to the average force on you multiplied by the collision time. The term we in physics use for momentum change is "impulse." That term also applies to the product of average force with collision time.

Think about that for a moment. When you jump off something and hit the ground, you can't alter your momentum change. While in the air, you are a projectile under the influence of gravity (and a little air resistance), which means the laws of physics dictate what your landing speed will be. Your mass is fixed and your velocity change is fixed. But your momentum change is equal to the product of two things, one of which you can control. By bending your knees upon making contact with the ground, you greatly extend the time it takes for your velocity to go to zero, compared with hitting the ground with your knees unbent. If you are able to extend the collision time by a factor of ten, you'll drop the average force on your feet, knees, and hips by a factor of ten.

Think about all the ways in which you have used or can use the bit of physics you just learned. Your car has one or more air bags in it. Which collision takes longer—your head on the steering wheel or your head on an inflated bag that deflates while coming to a stop? For less serious car collisions, your car's bumpers serve to extend collision times. Ever play catcher in Little League? That padded glove allows a baseball to come to rest in a longer time while compressing the padding compared to a regular glove. What if you don't have a glove and someone tosses you a ball moving a bit faster than you'd like? You catch it by moving your hand or hands back as the ball hits you. Pole vaulters land on a big, padded mat. Football and hockey players wear pads. Their helmets include padding, too, as do the helmets of cyclists.

Practice with Pads

Now think about Krav Maga class. We train on padded mats because if we are thrown to the ground, we don't want a short collision time with a hard floor. We use hand pads and body shields when a partner practices punches and kicks. In Figure 6.2 I am practicing an uppercut punch with my instructor, Mr. Abercrombie. Not only is Mr. Abercrombie using hand pads, I have on padded combat gloves. I'm also making sure to keep my guard up with my non-punching hand.

Now that you are armed with a little more physics knowledge, I'll add some quantitative details to the qualitative momentum concepts I've discussed. I put my combat gloves on and punched a force plate. I wasn't mad at the plate; I was punching it in the name of science. The goal was to study the force and collision time of my punch. I admit, however, that I didn't punch the force plate with everything I had. I was actually a little chicken because the thought of hitting a solid force

Figure 6.2
I hit a hand pad with my fist protected by a padded combat glove.

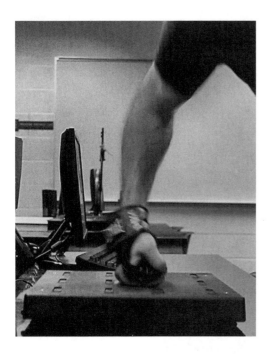

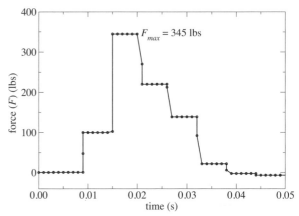

Figure 6.3

I punch a force plate in my department's physics lab.

Figure 6.4

A plot of the force of my punch on a force plate versus time.

plate that was sitting on a sturdy table didn't exactly sit well with me, despite the fact that I was wearing combat gloves. There isn't much "give" in the force plate and table, which means collision times are bound to be short, meaning forces are bound to be large. So, although I did hit the plate, there was no way I was going to hit the plate without a glove just so that I could demonstrate an even shorter collision time. Figure 6.3 shows my fist at the moment of maximum force on the force plate. My punch was downward, meaning it wasn't a typical punch, but that's not important here.

How long do you think the collision time was between my fist and the force plate? Analyzing the short video of my punch frame by frame, I found that I hit the plate with a speed of about 10.5 mph. That speed is about two-thirds of the speeds of the punches I analyzed in Chapter 3, but not too slow to be completely unrealistic for an actual punch that would be meant for someone's jaw. I certainly don't want to be hit with a fist moving over 10 mph! Figure 6.4 shows the force my fist exerted on the force plate during the collision as a function of time. The force plate is only capable of recording 1,000 data points per second and, like all measuring equipment, it has some error associated with it. That explains why the data are not smoother in the graph.

What Figure 6.4 shows you is that my punch's collision time was only something like 0.03 s. The blink of an eye takes 10 times longer than that. My fist's maximum force on the force plate reached 345 pounds, but keep in mind that Newton's Third Law says that that was the same size force my fist felt from the force plate. My combat gloves have less than half an inch of padding where my knuckles are. That is still enough padding to extend the collision time enough that I didn't hurt my hand.

But, "Wait," some of you more astute readers are saying. I told you in Chapter 3 that a force of 230 pounds could puncture skin. I can assure you that my fist was just fine after the punch in Figure 6.3; there were no skin punctures. Unlike the punch area I illustrated in Figure 3.3, the punch area I hit with in Figure 6.3 was easily twice as large. Increasing the area by a factor of two means the pressure drops by a factor of two. A couple of factors contributed to that area doubling. One, my padded combat glove allowed for a larger contact area upon punching. Two, I actually punched the force plate with more finger area than the knuckled area shown in Figure 3.3. I was simply too chicken to confine my punch to just my knuckles. I knew a little too much physics to punch the force plate like I would normally punch a hand pad.

Recall that impulse is change in momentum, and change in momentum is mass times change in velocity. My fist went from 10.5 mph to zero speed in 0.03 s. I can actually work out what mass is implied by the speed change, time of collision, and the force plot in Figure 6.4. My weight is about 190 pounds, which corresponds to a mass of about 86 kg. The mass of the moving object in my punch analysis came out to be 4.64 kg, which is about 5.4% of my total body mass. Is that my fist's mass? No way. My fist is only about 0.67% of my mass. What comprises about 5.4% of my body mass is my entire right arm. After all, it was my right arm that was moving during the punch. It's not like my fist was moving 10.5 mph at the moment of the punch while the rest of my arm wasn't moving. The beauty lurking in the laws of physics is that I calculated a mass corresponding to just the right percentage that my right arm occupies of my total body mass. I thus have confidence in my analysis and I gain a tiny bit more trust in the methods of science to accurately describe what is happening in the real world.

Now suppose I was masochistic enough to punch the force plate as hard as I could, but with no combat gloves on my hands. The lack of padding would reduce the collision time by a factor of two or three. That would increase the force on my hand by a factor of two or three, leading to a pressure large enough to puncture my skin. It's no wonder that people who punch walls while angry and drunk not only break a couple of fingers, they need a stitch or two on their bleeding knuckles.

You now know that padding extends collision time. My guess is that you'll see this concept all around like you never imagined. The next time you're at the amusement park and climb into a bumper car, I suspect if only for a second, you will think, "Good thing there is a bumper on this car to extend the collision times I'm about to experience!"

On to Fighting

The way we train in Krav Maga class for actual fighting is through sparring. There are many ways for people to spar, from fairly sedate, get-to-know-you friendly matches that introduce fighting, to all-out, full-contact sparring during which injuries can occur. Serious mixed martial arts (MMA) professionals may engage in the latter form of sparring. My daughters and I were introduced to the more friendly version while working on our black belts in karate. We wore chest protectors, shin guards, foot pads, helmets with face protection, and padded gloves. The goal in our karate sparring matches was to score points by successfully landing approved techniques on an opponent. Matches could certainly get intense, especially between more aggressive and gung ho students. But getting punched on my helmet or kicked in my chest protector never really hurt me. Padding did its job! Collision times were sufficiently extended so that the forces I felt when I got tagged weren't that large.

More than anything else, sparring helps acclimate us to facing off against someone who is trying to hit us. We not only want to defend ourselves, we want to put our opponent at ill ease by attacking and counterattacking. We get used to what it feels like to hit another person, and we get used to what it feels like to be hit. Safety is always of paramount concern, which means we spar in appropriate gear while being observed by professional martial arts instructors. Much like actual fights, sparring fights usually don't last very long. Once a person scores a point in our karate sparring, our instructor calls the point and there is a momentary stop in the action.

Our Krav Maga sparring is a step up from what I did in karate, but it's still nowhere near the full-contact sparring some MMA fighters engage in. We may spar for a minute or two without stopping, even if one person scores a hit on the other person. After the allotted sparring time, our instructor calls out the winner (there may be a tie). Our headgear does not come with face protection. We wear mouth guards to protect our teeth, but we are susceptible to being hit in small open areas on our face. We also wear 16-oz boxing gloves, which are heavier than the combat gloves we wear when hitting pads. There is more mass in a boxing glove, but the added padding protects us when we get hit, especially when we get hit in the head.

I asked in my first book[3] who are the best scientists? My answer is children. They *love* to ask questions, and they don't possess a fear of being wrong or "looking stupid," as some of my

3. *Gold Medal Physics: The Science of Sports* (Johns Hopkins University Press, 2010).

students unfortunately put it. Scientists have to ask questions and continually seek answers if we are to push humanity's understanding of the natural world to new heights. I often think of myself as a big kid doing a really fun job. But sometimes kids do stupid things and learn, one way or the other, not to do those stupid things again. I don't have a memory of sticking a key in an outlet before I turned one, but my parents assured me that I never did it again. At any rate, a question popped into my head one day while I was wearing my Krav Maga sparring gear. It was an admittedly stupid question, but one I couldn't get out of my head. So, like any scientist who is bugged by a question, I sought an answer.

That question is, What is it like to be punched in the face? Better yet, what can physics tell us about being punched in the face? I couldn't ask anyone to act as a guinea pig for me, so I decided to take one for science. I asked my former student, current physics department colleague, and fellow Krav Maga student, Dr. Moorman, to punch me in the face. Given that I'm one of her former college instructors, she readily accepted my offer.

I donned my sparring helmet and inserted my mouth guard. Dr. Moorman put on her 16-oz boxing gloves and prepared to punch me. A colleague of ours filmed me taking one for science. Figure 6.5 shows three photos from the video. The first gives you a sense of what my face looks like while not being punched. You can see the padding provided by my headgear, and you can see how much of my face is exposed. The second photo shows the moment of impact between Dr. Moorman's glove and my head. Her punch came in at 11 mph and hit me with a force of about 71 pounds. She definitely did not hit me with her most powerful punch, for which I'm grateful, but she didn't exactly pull her punch, either. My head felt her punch force as it experienced a maximum acceleration roughly four times the acceleration due to gravity (4 gees or 4 g). The third photo is my favorite. The skin and other loose parts of my face (lips, fat, and so on) have separated past my skull more than usual. My face looks like that of a baboon.

An important thing to notice in the second photo in Figure 6.5 is the slight compression of my headgear and the ever-so-slight compression of Dr. Moorman's boxing glove. Those compressions mean everything in sparring. They extend collision times, thus dropping the size of the forces we feel while being hit. It took just over 0.10 s for the collision between my head and Dr. Moorman's glove to essentially be over. That collision time was three or four times longer than the collision time when my combat-gloved hand hit a force plate in the previous section. There was no padding on the force plate and my combat gloves don't have as much padding as contained in the boxing gloves.

Warning: Please do not try getting punched in the face, and please do not take what I've written in this section as a suggestion that it would be a good idea to have someone film you while you get punched in the face. My headgear and Dr. Moorman's boxing glove helped prevent her punch from hurting me, but I was still at risk of sustaining a concussion, which is the most common type of traumatic brain injury. Take a good look at the third photo in Figure 6.5. Even with a moderate punch, you can see how my face reacts to acceleration. Once my skull has stopped moving, stuff that's loosely attached to my skull keeps moving until, like a spring, it's stretched to the point that the restoring force pulls it back. Our brains are not firmly attached to our skulls. They sit in a fluid called the cerebrospinal fluid. For very light impacts, meaning small accelerations, that fluid protects our brains like the pads I've discussed in this chapter. That's analogous to

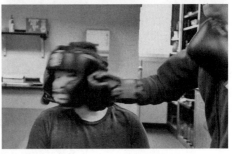

Figure 6.5
I'm taking a punch for science.

bumpers on cars. They help with low-speed collisions. But for large accelerations, just like in a high-speed crash between a car's front bumper and a tree, the cerebrospinal fluid isn't much help and gets pushed out of the way as the brain slams into the skull. You can think of what's happening to my face in Figure 6.5 as what's happening to my brain. I'm very lucky that Dr. Moorman didn't unleash a powerful punch to my head. My brain accelerated at 4 g, but that's nearly 20 times smaller than the acceleration that likely leads to a concussion. Some recent studies have concluded that concussions can happen for accelerations as low as 15 times what I experienced. I confess that my idea was a tad dangerous, because I didn't know how hard I was going to be punched, and I didn't know ahead of time what my brain's acceleration would be. There are two morals of this story. One, don't try this at home! Two, and more important, we spar while donning headgear and wearing gloves so that we don't come close to getting a concussion when we get punched in the head.

Time to Sprawl

What if you hear someone coming at you quickly from behind, and when you turn around, you don't have enough time to step aside? Using a person's linear momentum against him or her in the form of the flyby I discussed earlier in this chapter may not be an option because an attacker might be upon you so quickly that you have to react to certain contact. You can still make use of linear momentum ideas to help you with such an attack.

We learn "self-defense flow," which helps "develop our ability to respond to change in the dynamics of a physical altercation."[4] What I described in the previous paragraph is just such a "change." The basic moves we learn in beginning Krav Maga help us develop fundamental techniques that are used many

4. Warrior Krav Advanced Phase B Certification Manual.

times at the intermediate and advanced levels. But progressing to the advanced phases of Krav Maga means learning to adapt to changing situations. No attacker is going to tap you on the shoulder and let you have a few seconds to decide which beginning technique you wish to employ. An attacker might be an experienced fighter, and we should never underestimate the skill level of an opponent. After all, someone who risks an encounter with a potential victim does so without knowing the victim's skill level. Odds are, they take that risk with the awareness that a serious fight may result.

Okay, back to the situation I described, in which someone is running at you and you don't have time to get out of the way. Suppose, further, that your attacker is running low, in the hope of grabbing your legs and taking you to the ground. You most definitely don't want to be taken to the ground. In Figure 6.6, Mr. Abercrombie is running at me with the hopes of taking out my legs.

Figure 6.6
Mr. Abercrombie is about to take my legs out from under me.

What am I going to do? I've got my arms up, like most people would when seeing a person running toward them. Think about linear momentum. To change Mr. Abercrombie's linear momentum, I need to exert a force on him. He is put together quite well and he weighs just over 200 pounds. He may not be sprinting at me, but he's running at me fast enough that it will take a large amount of force on my part to stop him in his tracks. How large? Let's say he's running at me with a speed of 5 mph and I want to stop him in 0.5 s. I'm going to need to exert an average force of nearly 100 pounds to stop him. I'm capable of lifting 100 pounds, but think about where I'll need to generate a horizontal force that big. There better be plenty of friction between my shoes and the ground. I certainly couldn't stop Mr. Abercrombie if I were standing on ice. Recall also that my maximum force could be a factor of two or more larger than my average force. Not only am I not likely to stop Mr. Abercrombie, I'm likely to get hurt by the impact. If instead of going for my legs, Mr. Abercrombie tackled me at my center of mass and held on, we could move to the right in Figure 6.6 at a speed of about 2.5 mph.[5]

Recall that linear momentum is a *vector,* which means it has direction as well as a magnitude. The problem with the option I described in the previous paragraph is that I was trying to stop Mr. Abercrombie by exerting a force opposite the direction he was moving. That's the best way to bring him to a stop, but it's not my best option for handling such an attack. For all the force that's needed in the horizontal direction to stop Mr. Abercrombie, how much force is needed in the vertical direction to change the direction of Mr. Abercrombie's linear momentum? Because he's not starting with much vertical linear momentum, if any, I don't need to exert much force on him at all

5. I used conservation of linear momentum and took my weight to be comparable to Mr. Abercrombie's weight, though my body build is more like a pear.

to give him a vertical component of linear momentum. Now ask yourself which is easier, exerting an upward force on Mr. Abercrombie or a downward force? I'm going to choose the latter because gravity will help me. I may not be able to get out of Mr. Abercrombie's way, but I can make use of flyby ideas and at least get myself slightly off of Mr. Abercrombie's linear momentum direction. I can then further let Earth assist me in my defense by falling. A better and cooler sounding term than falling is "sprawling."

Check out Figure 6.7. Once Mr. Abercrombie is upon me, I anchor my arms on top of his upper back and drop. Gravity helps me exert the necessary downward force to alter his linear momentum direction such that he falls. The effort I need to exert is minimal, as is the amount of force needed to drop Mr. Abercrombie. That's another beautiful aspect of linear momentum. If it points in one direction, it doesn't take much to change its direction if there is a push perpendicular to its initial direction. Once I've sprawled and Mr. Abercrombie is on the ground, I quickly counterattack and then evacuate the danger zone as quickly as possible. If all goes well, the attack will be over in just a few seconds. Not only will my attacker's balance and equilibrium have been disrupted, I will have inflicted some damage with my counterattack, be it a punch to the temple or a kick to the kidney, the result being that my attacker won't know exactly in which direction I ran off.

As I mentioned before, sprawling is used as one part of an entire self-defense flow. That means that I may have defended against an initial attack, been pushed away while I tried to counterattack, and then felt the need to sprawl as my attacker executed his counterattack. There is no guarantee that my initial defense will work. Knowing a little physics makes a technique like a sprawl seem almost like an obvious reaction to someone trying to go for the legs of a potential victim. Fight back with physics!

Figure 6.7
I let gravity help me out by sprawling.

Gun Momentum

I'll end this chapter with a scarier scenario. It's similar to the ones I described in Chapters 2 and 3. Both those situations and the one here involve someone pointing a gun at you. In all cases, the attacker did something that gives you a great chance at escaping with your life. The attacker puts the gun on your body. You have the opportunity to get the gun's business end off your body before your assailant can react. If you have the opportunity to get your hands up, do so. Anything you can do to shorten the distance you have to move will cut valuable time off of your defense.

Look now at Figure 6.8. I have one of our training handguns on Mr. Abercrombie's back. I want you to imagine being in Mr. Abercrombie's position, but the loathsome individual with the gun is walking forward with you. In other words, you are being led somewhere. You don't know your attacker's intent and you really don't want to find out. Take advantage of your attacker's gun on your back and the fact that you and your attacker are walking forward. You are going to use your attacker's forward linear momentum against him or her.[6] Because your attacker and his or her gun have linear momentum in the forward direction, it would take a force in the backward direction to stop the motion of your attacker and his or her gun.

This self-defense technique was *made* for a physics class. Don't waste your attacker's forward momentum. If you suddenly disappeared, your attacker would keep walking forward, right? If you can move out of the way of the gun, your attacker will also keep moving forward. At some point during the terrifying situation you need to peek over a shoulder to determine which of your attacker's hands the gun is in. That dictates your rotation direction. Now look at Figure 6.9, in which Mr. Aber-

6. It's hard to convey forward movement with the photos I'll be showing you. I'd like to have made a video of the defense, but Mr. Abercrombie and I would have walked out of frame.

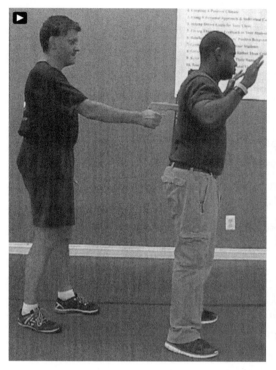

Figure 6.8
I have a training handgun at Mr. Abercrombie's back.

Figure 6.9
In the blink of an eye, my gun is off my target.

crombie has rotated counterclockwise (as seen from above). In just 0.45 s, Mr. Abercrombie has stopped walking forward by using friction between his right shoe and the ground, and he rotated toward me such that my gun's muzzle is no longer pointed at his body. If I've been preoccupied with moving my victim to some nefarious destination, chances are extremely high that I can't react in less than half a second and fire my gun. You can see the folly of putting a gun right on the victim's body. Please note that in Figure 6.9 I do not have my finger on the trigger of the training gun. That is a critical point in a training exercise!

Now think about linear momentum. My victim may have stopped on a dime and rotated toward me, but it's likely that I've taken a step forward while my victim initiated his self-defense move. You need to take advantage of that momentum. Though it's not so obvious in Figure 6.9 because Mr. Abercrom-

Figure 6.10
I'm in side control and
about to lose my gun.

bie and I weren't walking forward, were I a real attacker, my
forward momentum would have moved my gun closer to Mr.
Abercrombie's right shoulder than his left shoulder. Use the
linear momentum of your attacker to bring your attacker closer
to you. That gets your attacker's gun farther from where it can
hurt you, and you get to move in close—with the help of your
attacker—where you can turn the tables. Can you predict
what's coming next? You can almost see it in Figure 6.9

It took Mr. Abercrombie 0.6 s after the scene in Figure 6.9
to get to side control in Figure 6.10. That time would have been
reduced at least a tenth of a second had I been walking forward.
In a real situation, Mr. Abercrombie would have gone from victim to having turned the tables on his attacker in about one
second. You can see how important it was for Mr. Abercrombie
to utilize my linear momentum. He simultaneously got my gun
away from his body and let me walk right into side control. You

Figure 6.11
My right index finger is
about to be broken.

know what follows the scene in Figure 6.10. I'm about to be
pummeled with knee strikes and hammer fists and any other
combatives Mr. Abercrombie feels like unloading on me. I need
not show you all that punishment.

Let's move to Figure 6.11. Once I've been beaten a few times,
Mr. Abercrombie needs to remove my weapon. He keeps me
secure with his left arm as his right hand reaches for my gun
and he quickly rotates my gun clockwise (as seen by him). In a
real-world encounter, that likely breaks my right index finger.
This is why we never have our fingers on the triggers of our
training guns. To make matters worse—for me—once he has
my gun, Mr. Abercrombie strikes me with the muzzle, using the
gun as a cold weapon. I've been beaten, hit with my gun, and
now Mr. Abercrombie, in possession of my gun, is able to flee
the scene. I'll not show all that because I think you can visualize
it evolving from Figure 6.11.

The technique that I just discussed in concluding this chapter made use of linear momentum for only the blink of an eye, but that was the key to making the technique successful. I've discussed (and will continue to discuss) how the evolution of our bodies helps inform a martial arts system like Krav Maga. So do the laws of physics. The next time you get to a Krav Maga class, observe the techniques you learn through the lenses of physics and evolution. My guess is that at least one of them relies to some extent on making positive use of an opponent's linear momentum. Watch carefully because what I'm asking you to look for may happen quickly, as with the preceding gun defense. For some Krav Maga techniques, all that's required is a slight movement from an opponent to make all the difference between failure and success in the defense.

We Need More Power

7

What does the word "power" conjure up in your mind? Do you think of big muscles or a giant truck? Maybe you imagine a power plant that sends electricity to a town. But like many terms we've met so far, the word "power" in physics has a specific meaning.

Power is the rate at which energy is either used or outputted. I introduced this idea in Chapter 1. The great law in physics known as "conservation of energy" means that energy is never used up; it is only converted from one form to another. So it's the rate at which the conversion takes place that is of technical interest to the scientist and to the Krav Maga student. Even a spindly Krav Maga novice who is incapable of exerting much force may convert energy at a large rate. Training to have good fast-twitching muscles is incredibly important. Pace is also important because if you output too much power at the start of a fight, you might tire yourself out before you can safely get away.

The topic of energy conversions is of great interest in the modern world. Each of us in the United States, on average, consumes energy from the environment at a rate more than 100 times greater than what humans consumed a few hundred years ago. We love our gadgets that need to be charged each night, and we love our cars, refrigerators, microwave ovens, air conditioners, space heaters, and lava lamps. Energy conversions are continually happening all around us. Chemical energy stored in coal is released, which converts chemical energy in water to turn it from liquid form to gas form, which then converts to mechanical energy in spinning turbines, which then converts to electrical energy, which then may convert to heat energy in a toaster. None of our machines is 100% efficient, so some energy is lost with each energy conversion. That lost energy may go into heating a nearby stream or heating the atmosphere, both forms of pollution. In my previous discussion of energy (in Chapter 1) I focused on food energy, and that is the form that is most relevant to our examination of Krav Maga. Chemical energy in food is converted in our body to run all of the processes that keep us alive. But like all the energy examples I gave above, the chemical energy in food ultimately traces back to our sun. Energy in our sun ultimately traces back to the Big Bang. It's amazing to think about.

What we need to do now is convert some of the food energy we consumed into mechanical energy in the form of fast-moving elbows, fists, knees, feet, and heads. The better we eat and train, the quicker we facilitate the energy transfer. That means we have more power.

Potential Energy and Power

Any time an object gets stretched, it stores energy. The form that the energy takes is really chemical (and ultimately electrical) in that bonds between molecules and atoms are altered ever so slightly. But in the macroscopic world, we refer to any type of stored energy as *potential energy*. A muscle or tendon that's stretched will store energy. It's analogous to a stretched spring. We use little springs for our physics demonstrations. I have to do work to stretch a spring. The stored energy in the spring comes from me. We could go crazy and trace that energy back to the food I had for breakfast and the sunlight that helped my food grow and on and on. But if I do that every time I talk about potential energy, you'll pull out your hair.

Check out Figure 7.1. The top photo shows one of our small demonstration springs. Before I took the bottom photo, I

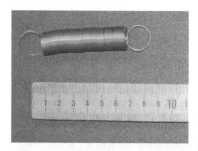

Figure 7.1

The top photo shows a magnified image of an unstretched spring. The bottom photo shows a stretched spring with 1 J of stored energy.

stretched the spring just far enough to have it store 1 J of potential energy.[1] Because my internal energy conversions are nowhere near 100% efficient, I might have done 4 J or 5 J of internal work to produce 1 J of external work in stretching the spring. "Joule" is a metric unit of energy and, as you know, metric units are not popular in the United States. I chose it here because it is readily used in formulas with another metric unit, watts (W), which is common in the United States. Because power is the rate at which energy is output per unit time, 1 W = 1 J/s. If it took me 1 s to stretch the spring, I would have output 1 W of power. If it took me 1 min to stretch the spring, I would have only output 1/60 W of power.

There is a neat way to determine how much power you can output. I do what I'm about to describe with my Physics of Sports class, and they always get a kick out of it. They challenge each other to see who can output 1 horsepower (hp),[2] which is pretty good. What you need to do is run up a flight or two of stairs. When you elevate your center of mass, you are increasing the gravitational potential energy you have with Earth. Measure the height of one stair and then count how many

---

1. Note that 1 joule (J) is roughly equivalent to 0.74 foot · pound (ft · 1b). I would have needed to stretch my spring about 89 cm (35 in) instead of 77 cm (30 in) to have 1 ft · 1b of energy stored in the spring.

2. 1 hp = 550 ft · lb/s $\simeq$ 746 W.

stairs you have to climb. Multiply that number by the height of a single stair and you'll have the total height you need to climb. Get a friend to time you as you run as fast as you can from the bottom of the stairs to the top. You also need to know your weight. *Warning:* The more flights of stairs you run up, the longer you will have to output energy. You'll certainly output more power running up a single flight of stairs than you will running up ten flights of stairs.

Take your weight in pounds, multiply that by the total height you climbed in feet, and then divide by the time it took you in seconds. Once you have that number, multiply it by 1.356 and you'll have your average power output in watts. It's an "average" power output because there is no way you could keep your power output constant while moving your legs up and down, swinging your arms back and forth, and turning around corners as the steps spiral upward. I ran up two flights of stairs leading to the back of our physics lab. There are 44 steps, each measuring 5.79 inches (as close as I could tell) for a total distance of 254.76 inches, or 21.23 feet. I ran up those stairs while fully clothed with shoes on, and with keys, wallet, and cell phone in my pockets. With all that added stuff, I weighed about 200 pounds when I ran up the two flights of stairs in 7.6 seconds. Following the recipe I gave you at the start of this paragraph, I outputted $(200 \times 21.23/7.6) \times 1.356$ W $\simeq$ 758 W $\simeq$ 1.02 hp. Yeah! I managed to get just over 1 hp, but I was panting for a minute or so afterward, and though I didn't test the following claim, I doubt I could have output 1 hp had I run up four flights of stairs instead of two.

Try what I just described. You'll have timing errors if you only run up two or three stairs. You'll be exhausted if you run all-out up 40 flights of stairs. Pick something reasonable, like the two flights of stairs I ran up. Many of you reading this book are in better shape than I am, so you could run up three or four flights of stairs without straining yourself too much. What I really want you to experience is outputting something like 1 hp. I'm able to match a draft horse, which is the animal for which the power unit was originally created, only for 10–20 seconds, depending on the time of day and how much I've eaten beforehand. You've heard a car ad telling you how much horsepower the car has. Do the experiment I described here and you can feel for yourself what it takes to output just one horsepower. It's time to put this book down and play with physics for a little while.

Elbow Power

Now it's time to get out of the back stairwell and onto the mat to discover the role that power plays in Krav Maga. When an attacker is up on you, too close for comfort, you don't have a lot of space to punch and kick. A great short-range technique is an elbow. Your arm is bent when you strike, so your arm's moment of inertia isn't as large as it is when you swing an outstretched arm for a roundhouse punch or a hammer fist. An elbow can be delivered in several different directions. For an assailant on your side, an elbow #2 is a combative that will help you deliver a quick strike with the hope of disrupting that person's balance. In an ideal world, you would like to step into your elbow strike, thus generating more power, but you may not have time to do that.

Figure 7.2 shows my instructor, Mr. Abercrombie, holding a body shield for me as I'm about to deliver an elbow #2 strike. My technique wasn't perfect, but what I was being trained to do was deliver an elbow strike with my weak side, and to do so quickly without stepping into the strike. We are trained by having someone bump us with the body shield, and then we look at our target and strike quickly. The idea is to prepare us for someone sneaking up on us or someone attacking us in a crowded room where we don't have space to move. I'm right-handed, but it's crucial that I train with both elbows. We never want to be in a position where we are attacked and then think, "Gee, I wish that guy had come at me on the other side, where

Figure 7.2
Mr. Abercrombie holds a
body shield while I prepare
to strike with an elbow #2
using my weak-side elbow.

I'm stronger and more confident with my striking ability."
That's a recipe for disaster.

My eyes are on my target. We should always be looking at
what we hit. My right hand is protecting my face in case my
attacker's mate comes at me from the other side. I've extended
my left arm so as to store potential energy in my stretched
muscles and tendons. I then pull my left arm to my left, bending my elbow and closing my left fist.

Figure 7.3 shows my elbow at 90° and moving at its maximum speed. I've converted potential energy into kinetic energy.
I've leaned into my target so that my center of mass moves
slightly to my left, thus increasing my elbow's kinetic energy
even more. My hips and core have rotated into my strike, which
also increases my elbow's kinetic energy. I need to develop more
core strength so that I can strike faster. My elbow was moving
over 13 mph. My right hand has dropped more than I would like.
But angular momentum conservation makes it difficult to ro-

Figure 7.3
My elbow has reached a maximum speed in excess of 13 mph. Note that my left hand is closed into a fist.

Figure 7.4
I yell upon striking, and Mr. Abercrombie feeds me the body shield to simulate hitting an actual person.

tate my left arm counterclockwise (as seen from above) without my right arm rotating clockwise (as seen from above). I did my best to keep my blind side covered, but I can't beat the laws of physics.

When I struck the body shield, Mr. Abercrombie fed me the pad so that I could experience resistance upon striking. That helps simulate hitting an actual person. Figure 7.4 shows my strike. My mouth is open because I yelled while striking. That helps with breathing and delivering more power. My left hand came up during the strike because Mr. Abercrombie pushed the body shield in a direction slightly off the horizontal. My guard on the right side of my face has dropped more than I would like, but that's why I train—to get better. I've mentioned this before, but it's worth repeating: if you can get a friend to film you while training, you can really help yourself improve if you watch the video after class. I pick out all kinds of flaws in what I do when I go frame by frame.

Because I can analyze my elbow strike frame by frame, I can determine the speed of my elbow as a function of time. Figure 7.5 shows just such a plot. Notice in the plot that my elbow moved around 5 mph while I was extending it to my right. That speed dropped just before I administered my elbow strike.

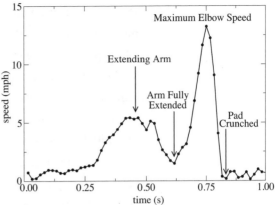

Figure 7.5
My elbow's speed as a function of time.

There are a lot of things I can calculate from the graph in Figure 7.5. My elbow slowed down upon hitting the body shield with an average acceleration of over seven times the acceleration due to gravity (7 g), but the maximum acceleration my elbow felt was about 11.5 g. I struck with a maximum force of about 131 pounds. My left arm's kinetic energy reached 90 J, which is nearly the kinetic energy of a baseball moving at 80 mph. My elbow's speed went from its maximum value to zero in just 0.083 s. That's a rate of energy dissipation into the body shield of nearly 1.1 kW ≃ 1.5 hp, which is the power of a typical microwave oven. My elbow strike rate of energy dissipation was 50% greater than what I could output running up the stairs in the example I gave in the previous section. I could reach 1.5 hp running up the stairs if, instead of running up the two flights in 7.6 s, I ran up them in 5.1 s. I have a lot of training ahead of me if I'm going to shave 2.5 s off my stair-running time. Keep in mind, too, that my average vertical speed while running up the stairs was only about 1.9 mph. That may seem slow, but remember that it took me just under 8 s to climb a bit more than 21 ft. I'm sure most of us could run 21 ft horizontally in much less time than 8 s. Also note that I must have had horizontal speed as well as vertical speed, so I was certainly moving faster than 1.9 mph. But I was not moving anywhere near as fast as my 13-mph elbow was before it hit the body shield. Kinetic energy scales with mass, but it also scales with the square of speed. I was moving a lot more mass up the stairs than I was when I threw an elbow, but my elbow was moving a lot faster than I was on the stairs, which is how my elbow power can be more than my climbing power.

Please don't confuse power with energy. The electric company charges us for energy used, not power. The company doesn't care how quickly we use the electric energy it supplies us, only the total amount of energy we use. All of us are capable of outputting large amounts of power—in the kW range—for very short periods of time. What really says something about your endurance is how long you can sustain a large power output. Energy output is found by multiplying average power output by time. The longer you output large power, the more energy you burn, which is why you get tired. From a Krav Maga standpoint, you want to be able to deliver energy quickly, meaning you can strike an opponent with a fast-moving elbow, fist, foot, and so on. It won't help your self-defense if you are capable of outputting the same amount of energy as your experienced instructor, but in twice the time needed by your instructor. Your chance of becoming a victim greatly increases if you are too slow, meaning you can't output energy very quickly, which in turn means your power output is too small.

Note from our discussion in Chapter 6 that the body shield extended the collision time between it and my elbow. Had I hit a concrete wall, my collision time would have been a great deal shorter, meaning the force on my elbow would have been a great deal larger. I would have dissipated much more power into the wall than I did into the body shield. The increased force and faster energy dissipation would have translated into a lot of pain for me.

Warrior Combinations

Knowing that some confrontations will require more than a single combative to end the fight, we are taught to link several combatives so that we develop a feeling for flowing from one move to another move. The idea isn't to memorize a sequence of moves and then hope you can employ that sequence without modification while being attacked. Real world fighting isn't that choreographed. But it helps to develop a feeling for why some combatives are linked together. We learn the "Warrior Combinations," about which we are told, "These are specific combatives put together to form various combinations."[3] There are 18 such combinations, but I'll focus on just one here. Warrior Combination #15 consists of three elbows with a spin, followed by three more elbows and then three knees. You may or may not think that a lot of exertion is required to complete that Warrior Combination, but if you have to repeat that sequence of combatives a few times in a row, you'll quickly appreciate how much exertion is required. Not only are we being trained to flow from one combative to another combative, we are training to output energy at a high rate. As a result, we develop more power.

Now let's get a little bit quantitative. I can execute Warrior Combination #15 in about 7.9 s, which is remarkably similar to the time it took me to run up two flights of stairs. I'm not going to show you all nine strikes and the spin in a sequence of photos, but I'll show you a sample of Warrior Combination #15 to give you an idea of what's involved. Figure 7.6 shows me in a ready position, about to perform Warrior Combination #15. It may not seem so, but there is a lot of physics in that photo. My stance is such that my shoes are just over shoulder width apart, thus ensuring a stable base from which to begin. My arms and hands are up in a defensive posture. By keeping my hands close to my face, I avoid slapping myself if someone strikes me from the side. Keeping my arms in close also allows me to move them quickly because I don't have to fight the larger moment of inertia that I would have had had my arms been at my side. As I'll discuss in more detail in the next chapter, a torque is needed to

---

3. *Warrior Krav Maga Phase A Certification Manual.*

Figure 7.6

I'm preparing to perform Warrior Combination #15.

get my arms rotating, and rotating an extended arm is tougher than a folded arm due to the former having larger moment of inertia compared to the latter. Mr. Abercrombie is not only training me to execute Warrior Combination #15 quickly and efficiently, he's providing—to the rest of the class—an excellent example of how to train a partner. Effective pad-holding techniques must be learned through training. Mr. Abercrombie has to put his hand pads in the appropriate places for me to hit. That means he must essentially go through the combination with me. His eyes are on me so that he can follow what I do.

My fastest-moving elbow happened to be the sixth and last elbow I threw, which makes me happy because I got faster as I moved through the technique. Figure 7.7 shows my right elbow moving forward at 19.8 mph. My left arm, which was used for the previous elbow strike, is tight against my body. That makes

Figure 7.7
My right elbow is moving
at nearly 20 mph.

it easier to rotate my body because my moment of inertia isn't
as large as it would have been had my left arm been flailing
outward. My eyes are on my target. I've developed good rotational speed by employing my core muscle group to rotate my
hips. My right leg is behind me, not only driving me forward,
but straightening out, which releases stored potential energy. I
feel like I did well on this particular strike.

Now it's time to dump a good chunk of my kinetic energy
into Mr. Abercrombie's hand pad. He's training me well because
he's moving the hand pad in his right hand forward. That allows
me to experience appreciable feedback during my strike, which
simulates hitting a person. The fact that his pad is moving in
the opposite direction as my right elbow means that a large
collision force is likely, which means my elbow could experience
a large acceleration as it slows on the pad. Figure 7.8 shows the

Figure 7.8
My right elbow feels 17 g of acceleration.

moment my right elbow experienced its largest magnitude of acceleration, which was 17 *g*. I certainly wouldn't want my *head* to feel that kind of acceleration. My entire right arm felt a maximum force of about 197 lb, but that was only for a brief period of time, something like 0.01 s. Forces are indeed large in order to create large accelerations, but those large forces don't last long. They do last long enough, however, that I wouldn't want to be hit by an elbow moving close to 20 mph. The time interval between Figure 7.7 and Figure 7.8 is only a bit more than 0.03 s, which should make it clear why accelerations and forces are large for an object moving nearly 20 mph that stops in just one-tenth of the time needed for an eye blink.

I'll now discuss one of my three knee strikes. It turns out that the third of my three knees strikes was my best, so I'll show that one. Figure 7.9 shows me in position to initiate a strike with my right knee. A knee strike should start with the striking leg extended back from the body. You will not generate

Figure 7.9
I'm about to deliver a knee strike as part of Warrior Combination #15.

much power if you strike with your front leg. It takes work to change kinetic energy, and work is the product of a force with a displacement. Not only do I have stored potential energy in hip flexors while my right leg is extended behind me; I have established a long distance over which to exert force on my right leg. The more work I can do on my leg, the more kinetic energy I've given my leg. Note in Figure 7.9 that I have grabbed Mr. Abercrombie's right shoulder with both my hands. Physics explains that, too. The force my knee exerts on my opponent will be larger if I pull my opponent into my knee instead of simply hitting a stationary opponent. The speed my knee moves relative to Mr. Abercrombie's chest is larger if I pull him toward me. Perhaps you know this from driving. It's much more dangerous to hit a car head-on that's moving toward you than it is to hit a stationary car. Relative speed is what counts. Always grab the shoulder opposite to your striking knee. If I had grabbed Mr. Abercrombie's left shoulder, my right knee might have only

Figure 7.10
My knee has reached 14 mph during Warrior Combination #15.

delivered a glancing blow. I'm aiming for his chest, either to knock the wind out of him or to break one or two of his ribs. I need a direct line to his chest, and grabbing his opposite shoulder gives me what I need. I also have the opportunity to store potential energy in my core muscle group, which I can release upon crunching my torso. On the training side, note Mr. Abercrombie's wide base and that he has double padded the striking area. My leg is three times more massive than my arm. It's better to be safe than sorry when holding pads.

My right leg reached maximum speed, shown in Figure 7.10, about 0.2 s after the image in Figure 7.9. As I noted, my leg is three times more massive than my arm, meaning it's harder to get my leg up to the same speed as my arm. My right knee was moving 14 mph in Figure 7.10, about 70% of the maximum speed my elbow had during one of my elbow strikes. Recall that kinetic energy scales with one power of mass and two powers of speed. Increasing mass by a factor of three while simultane-

Figure 7.11
My knee felt 12 g of
acceleration during
this strike.

ously dropping speed by 30% means kinetic energy increases by nearly 50%. When Mr. Abercrombie held pads for my elbow strikes, the pads were not on his body. My knee strikes were aimed at his body. Not only did my leg have more kinetic energy than my arm, my leg was heading toward Mr. Abercrombie's chest. There was thus an excellent reason for double padding.

The maximum magnitude of acceleration my right knee experienced was as it came to rest on the pads. I show in Figure 7.11 the moment when my right knee felt the greatest magnitude of acceleration, which was over 12 g. You can see how important it is to do lots of sit-ups and crunches. By pulling Mr. Abercrombie down so as to increase the relative speed between my right knee and his chest, I'm essentially doing a crunch while in a vertical position. I need to do many more crunches, because my torso should have been a little lower in Figure 7.11. Pulling an opponent into a punch, kick, knee, and so on, is another way to be tricky with the laws of physics. If the concept of relative speed is still a bit blurry in your mind, consider the following question. Would you rather be hit by a baseball thrown by someone capable of throwing 90 mph and who is standing in the back of a parked van or be hit by the baseball thrown by the same person standing in the back of the van, as it is driving away from you at 60 mph? You are standing on the ground, so what matters to you is the speed of the ball *relative to you and the ground*. In the first case, the ball moves 90 mph relative to you and the ground, but just 90 mph – 60 mph = 30 mph in the second case. You would definitely *not* pick the moving van option if the van was moving *toward* you at 60 mph. Aside from the danger of a van headed toward you, if the person in the back could manage to throw the ball with the same speed over the top of the van, you would see the ball coming at you with a speed of 90 mph + 60 mph = 150 mph. Air resistance and several other factors complicate my examples, but I hope you get the point. By pulling Mr. Abercrombie toward me, the speed with which his chest sees my right knee moving increases.

My knee might have felt 70% of the maximum magnitude acceleration that my elbow felt (12 *g* compared to 17 *g*), but my leg is three times more massive than my arm. That is why my leg delivered a force about twice what my arm delivered. Remember that force scales with one power of mass and one power of acceleration. The 400-lb estimate I made for the maximum magnitude force felt by my right leg is a rather rough estimate. A lot more than my leg is moving. Mr. Abercrombie is moving. I in fact managed to pull his chest toward me at about 2.8 mph relative to the ground at the moment my knee struck the pad. The numbers here are not meant to be exceedingly precise. They are, however, reasonable estimates for the sizes of the speeds, accelerations, and forces involved in the elbow and knee strikes delivered during Warrior Combination #15. They also give you a good idea why a knee strike could deliver about twice as much force as an elbow strike. That doesn't mean, of course, that a knee strike is twice as effective as an elbow strike. Effectiveness doesn't just scale with force. The part of the body being hit is a major player in deciding which strike causes more pain. If I land an elbow in your eye, you might claim that that strike is much more painful than if my knee strike had caught you in your gut instead of on your sternum. You might have great abdominal muscles and weren't hurt much with my knee, but if I popped your eye with my elbow, you won't care how hard my knee hit you.

Even though it took me only about 8 s to complete Warrior Combination #15, I was a bit winded after doing it. Delivering three elbows, followed by a spin, then three more elbows, and finishing with three knee strikes is hard work. Pulling that work off in a short time means outputting a lot of power. Warrior conditioning means performing two or three Warrior Combination #15s in a row. It means being able to sustain a fight long enough to deliver plenty of punishment while leaving yourself able to flee the scene once the punishment ends. Technique is important and if, say, Warrior Combinations are done improperly, one might be wasting energy with wasted movements. The evolution of Krav Maga techniques has tried to eliminate movements that waste valuable energy. If you have to expend a great deal of energy to defend yourself against an aggressive and/or talented attacker, you'd better be conditioned for the encounter.

Watch Your Side

One possible dangerous situation involves a person running at you. Options for moving aside and letting the assailant fly past may not exist. Your best option might be a defensive side kick. An even scarier situation is one in which you are engaged with one attacker and out of the corner of your eye you see another attacker. You don't have time to use the person you're fighting as a shield, so you opt for a side kick to ward off the second at-

Figure 7.12

Mr. Abercrombie is about to kick me.

tacker. If two people are engaging you, an offensive side kick may be your best option. Such a kick, if landed well, may seriously hurt someone and give you a chance at escape or at least the chance to fight one person at a time.

A side kick is a powerful weapon in your arsenal. Though everyone has a unique build, an average human adult's leg is about one-sixth of his or her body weight. If you weigh 180 lb, that means your leg weighs about 30 lb, or roughly the weight of two bowling balls. Imagine two bowling balls attached to each other and swinging at you. That comparison breaks down in a few different ways, because your leg's mass is distributed over a larger volume, bowling balls are significantly harder than the flesh on your legs, and your legs have joints like ankles and knees that are capable of bending portions of your legs.

Figure 7.12 shows me holding a body shield while Mr. Abercrombie is about to unleash side-kick fury. Note Mr. Abercrombie's eyes are on his target. Always look at your target. A very quick glance can mean the difference between an effective hit on your attacker and injuring yourself while wasting a kick. Mr. Ab-

Figure 7.13
Mr. Abercrombie's back leg
has slid up to his front leg.

ercrombie is looking over his right shoulder. By turning his torso counterclockwise (as seen from above) he is storing potential energy in his core muscle group. Note also how I am holding the body shield. I have bent my knees and assumed a wide stance. Mr. Abercrombie may be kicking, but I am training, too. I am getting used to absorbing a kick. My wide base ensures good balance and makes it difficult for Mr. Abercrombie to knock me over. I'm going to need that balance, as you'll soon see. My body shield is pressed tightly against me so that it doesn't slap into me upon being kicked. A pad holder has a lot to think about.

To execute a great side kick, make use of as many physics principles as possible. The attacker is going to feel the kick, so you might as well have a little forward momentum while you kick. Recall that linear momentum is the product of mass and velocity. Having one-sixth of your body's mass in your leg, and then putting that mass in motion, will make for some significant linear momentum. Figure 7.13 shows that Mr. Abercrombie slid his left leg up to this right leg. He has also pivoted his left foot so that he can maintain balance during the time of the

Figure 7.14

Mr. Abercrombie's right leg is cocked for a side kick.

strike. He knows the line of force with his kick will be joining the two of us, so a pivoted foot will keep him from tilting to one side. Mr. Abercrombie is a fifth-degree black belt in karate, so he's had a lot of practice executing side kicks. Also note that Mr. Abercrombie's eyes are still dead on his target.

Now it's time to store some potential energy, but only for a fraction of a second. By cocking his striking leg, Mr. Abercrombie stores energy in the powerful muscles, tendons, and ligaments in his leg and his core. Figure 7.14 shows a cocked right leg in preparation for a side kick. The time elapsed between Figure 7.13 and Figure 7.14 is only about 0.12 s. Note Mr. Abercrombie's arms are not compressed on his body, but out slightly. His torso has rotated toward the right as his right leg moves to the left. That redistribution of mass ensures balance. Eyes are still on his target.

Now comes the strike. Roughly 0.13 s after Figure 7.14, full impact from the side kick arrives in Figure 7.15. Mr. Abercrombie's foot was moving at 13.6 mph (21.9 kph) when it first made contact with the pad. What I show in Figure 7.15 is the pad fully

Figure 7.15

Full impact of a powerful side kick.

compressed while I am fully braced for the impact. Be sure to note that Mr. Abercrombie's right knee is bent. We discussed this in Chapter 3 with punches. Maximum speed occurs when the leg is bent, not when the leg is fully outstretched. That fact must be considered when estimating target distance necessary for an effective side kick. Besides keeping his eyes on his target, Mr. Abercrombie's left arm has moved farther from his body. That is because his right leg is extended farther in Figure 7.15 than in Figure 7.14. As I discussed in Chapter 2 and will describe in more detail in Chapter 8, extending his right leg means Mr. Abercrombie is more prone to rotate forward. Rotating his torso to the right and moving his left arm farther back to the right, as seen in Figure 7.15, provides backward rotation to counter the forward rotation from his extended right leg. Balance is again the goal. Note, finally, that Mr. Abercrombie came up off his pivoted left heel. He had a little more desire for power when I challenged him to "hurt me" with his side kick!

After having delivered a powerful side kick, Mr. Abercrombie's right foot returns to the mat, his arms are in tight, and his

Figure 7.16
Damage has been inflicted.

eyes remain fixed on the target he just demolished. Figure 7.16 shows a wrinkled body shield and me having recoiled from a powerful side kick. The keys to the power in the kick are cocking the leg prior to executing the kick, proper rotation and balance, and converting stored energy into kinetic energy in a short amount of time. Cocking the leg ensures that core and leg muscles act on the leg over a large distance, thus performing more work on the leg, which in turn leads to greater kinetic energy. The time elapsed from Figure 7.12 to Figure 7.16 was a whopping 0.95 s, which means energy conversions happened in short time scales, which further means Mr. Abercrombie delivered a lot of power in his side kick.

Think about everything I just described in delivering a powerful side kick. Mr. Abercrombie acquired a target, executed his kick, and returned to a defensive position in just under a second. For an additional attacker coming to help a mate attack you, freeing yourself from the first attacker for just a second can be enough time to deliver a powerful side kick to that second attacker. The kick may also be delivered if your arms are engaged with the first attacker. We have four limbs. Use them as much as possible while in an actual confrontation.

Ouch, My Knee

Figure 7.17 shows Ms. Maupin using an angle kick to hit a pad held by me. From the moment of first contact with the pad to her leg briefly coming to rest at full compression of the pad, about 0.1 s elapsed. Her average force on the pad during that time interval was nearly 350 pounds. Generating a lot of force on a pad like that requires a body that is capable of converting chemical food energy into mechanical leg energy very quickly. Also needed is a good technique that does not waste energy in movements that will not contribute to a hard hit. It should not be necessary to show other photos, pre- and post-hit, of Ms. Maupin's angle kick strike. You already know enough physics

Figure 7.17
Torque the hip to get lots of power.

that you know Ms. Maupin's initial position consisted of a well-balanced stance. After her leg rebounded from the pad, she returned to a well-balanced stance. You can see in Figure 7.17 that, as the pad holder, I have a wide stance. That stance is oriented such that the line joining my shoes happens to be the line of force from Ms. Maupin's kick.

I'll close this chapter with one very important safety tip concerning what you see in Figure 7.17. Note that my forward left leg has been turned. You can tell from my left shoe how my left leg is oriented at the time of Ms. Maupin's strike. The body shield I was holding was resting on the *side* of my knee. It was also touching parts of my leg above and below my knee. I wanted the kick force to be distributed along the side of my left leg. That greater surface area meant smaller pressure on the various parts of my leg that felt the force from the pad. Think about what would have happened had my foot been pointing toward the camera that was filming the video of the kick. My bent and pointed knee would have been touching the back of the pad. Because the pad doesn't easily bend, little else of my leg would have been touching the pad. Not only would Ms. Maupin's kick force have been concentrated on my knee, the line of that force would have caused my knee to bend in a direction opposite to that provided by evolution. But that's the entire point of the angle kick. If you aim such a kick at an assailant's knee, you have a great opportunity to shatter that person's knee as you not only concentrate your kick force on the knee, you force the knee to move in a direction it doesn't want to move. That's what makes the kick so effective; it exploits how our knees evolved. But that's why we have to be careful during training. We don't want the reason the kick exists in the Krav Maga repertoire to hurt the pad holder. So, please, if you're holding a large pad for a mate to practice an angle kick, think about physics and hold the pad in the correct orientation. I don't want any of my readers or their mates to have shattered kneecaps.

Rotate Your Way to Safety

Rotations in Krav Maga are where some great physics is spinning around. Newton's Second Law for rotations and angular momentum conservation help us understand all of the rotational movements in martial arts systems. Like so many of the topics discussed in this book, I think you'll learn a little physics here, and then start seeing physics in action all around you. Let's get down some basics of rotational physics before stepping back into Krav Maga class. Some of what I'm about to discuss has been mentioned earlier, but it won't hurt to add a few more mental repetitions to your physics training.

8 CHAPTER

Rotation Basics

Think of all the things around you associated with rotation. Relative to our sun, Earth spins on its axis once every 24 hours. That is how we define a "day." It's Earth's essentially constant spinning that dictates when you need to go to work, when you need to go to sleep, and when it's time for Krav Maga class. If someone could hop off Earth at the equator, suspend himself or herself in space, and watch Earth turn, that person would see objects at the equator fly by at over 1,000 mph. Lucky for us that we have a strong gravitational pull from Earth to keep us safely on Earth's surface. We define a year by the time it takes Earth to make one compete revolution around our sun. To travel such a distance in a year, Earth averages about 18.5 miles per second (more than 66,000 mph). We don't notice that seemingly fast speed any more than we notice going 500 mph in an airplane. Earth is our reference frame. And in our reference frame, we watch the moon orbit our planet in about a month.

What other spins do you notice? Any of you reading this book recall when clocks had hands? Recall record players and compact discs? Just about all of us have used computers with spinning disk drives. Car wheels turn, as do washing machines and dryers. I must rotate the door I walk through when entering *Warrior Success Academy* in Forest, Virginia for my Krav Maga class.[1] In the sports world, baseballs spin, as well as soccer balls kicked so their trajectories are bent like David Beckham used to bend them. Well-thrown footballs have 600-rpm spirals and some brutal tackles rotate ball carriers into next week. I could go on and on, filling scores of pages with nothing but examples of rotations we encounter on even the most mundane of days, but it's time to get to the physics.

I talked about torque in Chapters 2 and 5, but it's worth reminding you that a torque is the rotational analog of force. What's added with torque is a lever-arm distance. Multiply force and lever-arm distance and you get torque. Just as a net, external force is required to change an object's linear motion, a net, external torque is required to change an object's rotational motion. An object spinning at a constant rate may have several torques acting on it, but it has no *net* torque on it. And just as a net force is needed to change an object's linear momentum, a net torque changes an object's *angular momentum*. To continue the analog theme, just as linear momentum is the product of mass and velocity, you can think of angular momentum as the product of "moment of inertia" and "angular velocity." As for directions, I'll use clockwise and counterclockwise (as seen from a given point of view) and hope that readers have seen a clock with hands.[2]

What I wrote in the previous paragraph summarizes two chapters in an introductory physics book. But we're good to

---

1. After doing so, like all students, I bow and say, "Hello sir. Hello ma'am." That is our way of offering respect upon entering.

2. Physicists don't use clockwise and counterclockwise. We have more formal, mathematical ways of defining directions when it comes to rotational quantities, but those methods are outside the scope of what we need for Krav Maga physics.

proceed with the conceptual definitions I just gave. Besides, I've already thrown those rotation physics terms at you many times in this book. I'm trying to build your intuitive physics, and repetition is crucial. One key point to keep in mind is that any object rotating, or moving along a circular path, requires at least a force pointing in toward the center of the circle. That force is often called the *centripetal* force. Nothing can move along a circular path without something pulling or pushing it toward the center of the circle. A car rounding a turn is a perfect example of an object whose velocity is changing because the car's direction changes. Even if the car rounds the turn at constant speed, its velocity changes. That change comes about because of a net, external force. Something has to keep the car from moving along a straight line. That something is a force from

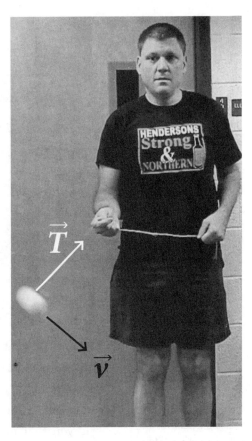

the road, either in the form of friction or a normal force from a banked road.

Probably the simplest example I can imagine is twirling a ball in a circle that is tied to a string. See Figure 8.1 for a photo of my simple demonstration. This demo might be easier to understand than reverting back to Figure 1.1 and talking about me pulling inward on a jump rope. The string's tension provides an inward force that keeps the ball moving in a circle. The ball only weighs 0.875 oz, so I don't need much force to keep it moving in a circle. The most tension I provided in Figure 8.1 was about half a pound. I need to include the ball's weight when analyzing the motion, but the point is that an inward force is needed to keep the ball moving in a circle. I show my tension force ($\vec{T}$) and the direction of the ball's velocity ($\vec{v}$). If the ball is to speed up or slow down, there must also be a force along the direction of the

Figure 8.1

I'm spinning a whiffle ball attached to a string. I'm doing my best to maintain a constant speed for the ball, but that doesn't remove the need for an inward force to keep the ball moving in a circle. My tension force ($\vec{T}$) and the ball's velocity ($\vec{v}$) are shown. The ball is moving counterclockwise.

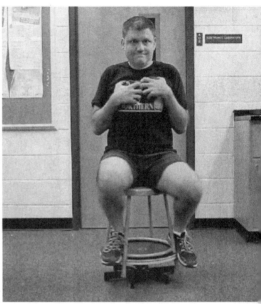

Figure 8.2

My outstretched arms keep my moment of inertia large; pulling in my arms reduces my moment of inertia. Note that in the photo on the right, I am slightly blurrier than I am in the photo on the left, which indicates that I'm spinning faster on the right.

ball's velocity (or opposite, if slowing down), but that doesn't change the need for an inward component of force.

I'm going to walk you through one more demonstration of a rotational concept before moving on to Krav Maga physics. I told you that moment of inertia is the rotational analog of mass. Just as you need a lot of force to alter a really massive object's state of motion, you need a lot of torque to alter the rotational motion of an object with a large moment of inertia. If I happen to be analyzing a system of objects for which the net, external torque is small compared to any of the torques between objects in my system, then that system's angular momentum is essentially conserved. Because angular momentum is the product of moment of inertia and angular velocity, doing something to make moment of inertia big will make angular speed small, and vice versa. For the demonstration photographed in Figure 8.2, I sat in a swivel chair with low friction, which means there was little external torque on the chair and me. I got myself spinning while holding some weights in my hands. Because my arms were outstretched, my moment of inertia was relatively large. The more mass I move away from a rotation axis, the larger the moment of inertia. When I pulled in my arms, thus bringing more mass closer to the rotation axis and lowering my moment of inertia, I rotated faster. This is something you can try on an office chair. Holding masses (books, bottles, or whatever) in your hands helps make the reduction in moment of inertia more dramatic.

My rotational speed in Figure 8.2 went from 27 rpm to 46 rpm. Because my angular speed increased by a factor of 1.7, my moment of inertia dropped by a factor of 1.7 when I pulled in my arms. You've seen this phenomenon before when you've watched figure skating. As a skater enters her final spin, she starts with both arms extended outward and usually one leg extended outward. That makes her moment of inertia as large as possible. She then pulls everything in as close as possible to her spin axis, which makes her spin so fast that she's nothing but a blur. That's angular momentum conservation on full display.

An important concept to remember with my chair demonstration is that pulling mass toward a rotation axis lowers moment of inertia and makes rotational speed go up. It's also easier to rotate a stationary object if its moment of inertia is small, in the same way it's easier to accelerate an object if its mass is small. We're now ready to rotate back to Krav Maga class.

Inflicting Pain Up Close

Let's play with the idea that rotating an object with small moment of inertia is easier than rotating an object with large moment of inertia. When fighting in close quarters with an opponent, we often don't have the space to execute a side kick or throw a jab. There are times when an elbow or knee is a great option. But there is more to an elbow or knee technique than simply being space limited. In-close fighting is likely not something that can be sustained for very long. You must inflict pain and separate from your attacker quickly. We don't want to

Figure 8.3

Mr. Abercrombie uses hand pads to help train me as I throw an elbow (*left*) and a knee (*right*).

be stuck fighting with some guy while his mates have time to add to the threat.

You know from the previous section that an outstretched arm or extended leg will have more moment of inertia than a tucked arm or tucked leg. For the same angular acceleration, it thus takes less torque to swing an elbow or knee than it does to swing an extended arm or extended leg. If we choose to employ the same torque from our skeletomuscular system, we'll get more angular acceleration out of an elbow or knee than we could get from an outstretched arm or extended leg. Check out Figure 8.3, which shows me practicing an elbow and a knee with my instructor, Mr. Abercrombie. I discussed elbow and knee strikes in Chapter 7, but in this chapter, I want to focus more on the rotational aspects of the techniques.

Though my technique still needs improvement, note that when I throw an elbow or knee, my entire body is involved. I generate a large torque by using my core to rotate my entire torso while throwing an elbow. My hips rotate, and I use the friction between my shoes and the mat to stabilize me as I experience a backward force from the hand pad as I hit it (recall Newton's Third Law). When I throw a knee, I use my core to crunch down my torso while pulling up my knee. That helps stabilize me, because my knee rotates counterclockwise while my torso rotates clockwise. Note, too, how I pulled Mr. Abercrombie down as I delivered my knee strike. That increased the force on the hand pad, because the hand pad moved toward my knee as my knee struck. I'll generate even more torque with my elbow and knee as I improve my core strength.

Because my arm and my leg are bent in Figure 8.3, I get close to a factor-of-two reduction in moment of inertia for my arm and my leg, similar to the reduction I got when I pulled my arms inward in Figure 8.2. That nearly doubles the rotational speed with which I can deliver an elbow or knee, when compared with swinging a straight arm or straight leg. I've got a lot of work left to perfect my techniques, but I can tell from the hand-pad depressions in Figure 8.3 that I'm performing well enough to inflict serious pain if the hand pads were replaced with vulnerable areas on an attacker. The great thing about the physics concepts revealed in this section is that they allow *anyone* to throw an effective elbow or knee. Even the most diminutive person can generate a lot of rotational speed with an elbow or knee. And if that person's elbow lands on an attacker's nose or if that person's knee lands on an attacker's groin, that small person has a great chance of escaping danger.

Rotate Out of That Headlock

There are plenty of reasons for worry if someone messes with our heads. An example I discussed in Chapter 2 is how getting poked in the eye causes us angst. But there are many other reasons to be concerned if our head is the object of an attack. Our trachea runs up the front of our neck, making us vulnerable to an attack that could disrupt our breathing. A normal adult takes something like 15 breaths per minute. That translates to one breath every four seconds or so. Though we can consciously increase or decrease our respiratory rate, we usually take breaths without even realizing we're doing so. Imagine taking a breath, holding it, and then counting off how many seconds you can go before you really want to take another breath. Some of us can swim under water for as long as a minute (or more), but we're anxious to breach the surface and take a big breath. If, unlike our voluntary suspension of breathing as we swim underwater, an attacker does something to hinder our breathing, we are easily susceptible to panic. None of us wants to go long without taking a breath. While under stress, our respiratory rate increases and we are even more anxious for each breath.

Suppose an attacker surprises you from the side and puts you in a headlock. It's possible your breathing will be compromised. There are other dangers, too. Your carotid arteries run up each side of your neck, supplying oxygen-rich blood to your brain. Our brains don't store energy, so we can't talk about a great reserve of potential energy nestled in our brains that we can utilize if the blood supply is diminished or stopped. If our blood supply is cut, we are in serious trouble. Getting choked by an attacker has the potential to knock us out. It takes less than 10 seconds of oxygen deprivation for our brains to shut down so that they can conserve what little energy they have. We definitely don't want to become unconscious because we didn't correctly guess the intent of our attacker. So breathing and keeping our brains functional are of paramount importance when someone tries to choke us.

How do we train without causing each other harm? We must train with safety in mind, but we have to provide a reasonable simulation of getting attacked so that we are prepared for an actual assault. Headlocks are lightly applied so that we aren't actually choking each other. In Figure 8.4, Mr. Hall puts Mr. Abercrombie into a side headlock. Note that Mr. Hall isn't trying to seriously choke Mr. Abercrombie. The self-defense technique that Mr. Abercrombie is going to demonstrate is that

of the "early interception of attack."[3] He's going to react as soon as he feels Mr. Hall applying a side headlock. The other technique that we practice is "late when the headlock is on (start with eyes closed)."[4] As the name implies, the attacker already has us in a headlock, and we need to get out of what could potentially be a very tight grip. But I want to discuss the physics behind the quick-reflexes response.

In an actual confrontation, Mr. Abercrombie would need to create space between his neck and his attacker's arm. Space already exists during training, but it's good to go through the motions that would be necessary in a real attack. The first thing Mr. Abercrombie does is reach for Mr. Hall's arm. But along the way, he uses his fingers to pluck the eyes of his attacker. This is only simulated while training. Figure 8.5 shows Mr. Abercrombie's hands flying up toward Mr. Hall's face. The idea is to exe-

Figure 8.4
Mr. Hall puts Mr. Abercrombie into a "light" side headlock.

Figure 8.5
Pluck those eyes—but while training, not really.

3. Warrior Krav Level 1 Certification Manual.
4. Warrior Krav Level 1 Certification Manual.

Figure 8.6
Rotation has begun. Mr. Abercrombie's right hand has created space between his neck and his attacker's arm by pulling his attacker's arm after an eye pluck.

Figure 8.7
Heading to side control.

cute that motion at the moment you feel an attacker's arm go to your throat.

Now we come to the all-important rotation. Mr. Abercrombie doesn't want to move his head forward or he risks choking himself. He needs to rotate away from his attacker's threatening arm. Check out Figure 8.6, and note that Mr. Abercrombie has securely grabbed Mr. Hall's arm. As seen from above, Mr. Abercrombie is rotating counterclockwise, meaning his neck is rotating toward his attacker. Note that Mr. Abercrombie's arms are in tight, which keeps his moment of inertia down and allows him to rotate faster.

The next step is key, because you never want to cede control back to your attacker. Figure 8.7 shows Mr. Abercrombie rotated slightly more than he was in Figure 8.6. As Mr. Abercrombie continues to rotate, he swaps his right hand for his left hand on Mr. Hall's arm. Mr. Abercrombie's right arm is shooting upward in preparation for side control. The goal is to do the sequence of moves smoothly and quickly. Achieving that goal takes lots of practice. Don't worry if it takes a long time to master the hand transfer. We are taught a variety of responses if we execute a technique poorly. If you had to get out of a side headlock and lost control of your attacker's arm while rotating into side control, just keep fighting. You may need to use a completely different self-defense if your attacker manages to get away from your initial self-defense technique. Actual fighting will never be as clean as what you practice in the safe environment of a Krav Maga class.

By the time the technique has progressed to Figure 8.8, Mr. Abercrombie has gained the upper hand after getting Mr. Hall into side control. Only about 1.4 s have elapsed between Figure 8.4 and Figure 8.8. Note that Mr. Abercrombie's right knee is on the way up as he's pulling Mr. Hall down. Deliver combatives as soon as you put your attacker in side control. Don't let precious seconds slip away; you don't want your attacker regaining balance and regaining the upper hand. The key points in this self-defense are creating space between an attacker's arm and your neck, followed by an efficient way of getting out of the danger zone. By keeping your arms in tight, a rotation into your attacker can be executed quickly. It may seem obvious that you would prefer facing your attacker instead of having your back to your attacker. And it may seem equally obvious that if you aren't facing your attacker, you'll want to rotate in such a way that you can face that person. What physics helps us understand is why Krav Maga techniques have evolved in the way they have. Having your mass near a rotation axis helps you rotate quickly. It also helps with close-quarters counterattacks because arms and hands are right where they need to be for moves like side control.

Figure 8.8
Punishment time in side control.

Rotate That Knife

Sometimes the rotations we need up close are for defenses in more harrowing situations, such as an attacker wielding a knife. Figure 8.9 shows Mr. Abercrombie about to attack me with a training knife. He is about to use a roundhouse motion with his right arm, perhaps trying to stab me in my left side. Were that a real attack, doing nothing to defend myself means risking a stab to my lung, or perhaps my kidney if my attacker stabs around my back. In Figure 8.9, I'm playing the victim, but doing my best to demonstrate my training. My eyes are on Mr. Abercrombie's knife and, as we're taught about a knife fight, I'm trying to prepare my mind for the likelihood of being cut. I also need to be on the lookout for possible additional weapons that my attacker might have.

The attack begins in Figure 8.10. I have rotated my left arm outward to stop Mr. Abercrombie's knife from hitting my left side. But as I've discussed a few times now, I don't try to hit his

Figure 8.9

Mr. Abercrombie is preparing to attack me with a training knife. He will try to stab me with the knife held in his right hand and using a mid-height roundhouse stab.

right forearm with the intention of bringing his arm's speed to zero. That takes too much force. I have instead bent my arm so as to reduce its moment of inertia, which allows me to rotate it faster. I'm going to keep rotating my arm, which will rotate his arm. I have changed the direction of his arm's linear momentum, but have not eliminated it. Mr. Abercrombie is good enough with knives that he has instinctively turned the knife in his right hand outward, transforming his intention from a stab to a slice. That is a reflex I will need to learn. The idea is that his initial attack failed, so he's adapting to the new configuration of his arm by preparing a slice counterattack. Only 0.7 s have elapsed between Figure 8.9 and Figure 8.10. Given such a short time that I had to react to my attacker's knife thrust, it's a good thing I opted for an arm rotation with small moment of inertia.

Now look at Figure 8.11, which is just 0.17 s later. I have moved toward my attacker, extended my left arm slightly, and I am in the process of rotating Mr. Abercrombie's knife-wielding

Figure 8.10
The attack begins. Just 0.7 s have elapsed since the scene in Figure 8.9.

Figure 8.11
Keep the rotation going.

Figure 8.12
My arm rotation has reached 45 rpm.

Figure 8.13
I'm nearly in a position to inflict some revenge on my attacker.

arm. Remember, once my arm is in rotation, it will take a torque to slow it down. I might as well keep rotating. That is a key physics concept in Krav Maga. Parry or deflect an attacker's arm or leg so that little force is exerted in avoiding being hit or cut. Once we are in counterattack mode, keep that momentum going, again to avoid the excess force needed to stop it. Use momentum, be it linear or angular, to your attacker's disadvantage and to your advantage.

In Figure 8.12, 0.33 s after Figure 8.11, I have rotated Mr. Abercrombie's arm toward the vertical. I've moved in a little closer, and my right hand can just barely be seen getting ready to put Mr. Abercrombie in side control. The rotation rate of my arm is about 45 rpm, one-tenth of the rotational speed of a helicopter's rotor blades. I need to develop a faster motion because my attacker could be skilled enough to move faster than me.

Just 0.2 s later, Figure 8.13 shows that as I am in the process of completing my rotation, I'm moving forward, which closes the space between me and my attacker, and my right

arm is rising as I prepare to put him in side control. Note that my right arm is not extended, because I not only want to keep my arm out of the way of the swinging knife, I also want a small moment of inertia so that I can get my right arm in position quickly.

Now check out Figure 8.14. In the 0.33 s since the scene in Figure 8.13, I've closed the gap between my attacker and me. My right arm is shooting over Mr. Abercrombie's shoulder so that I can get it behind his head for side control. My left arm has extended, and will soon work with my right arm in side control. The extension of my right arm serves another purpose. That arm's moment of inertia has increased, which makes it more difficult for Mr. Abercrombie to fight its rotation. Moment of inertia plus linear and angular momentum are going wild as they help me in every way they can.

Figure 8.15 shows the end of my first defense against my attacker's knife stab. I have Mr. Abercrombie in side control, including pinning his right elbow hard against my shoulder. That

Figure 8.14
Almost ready for side control.

Figure 8.15
I've got my attacker in side control and I'm ready to deliver punishment in the form of combatives.

Figure 8.16
Mr. Abercrombie maintains control of my hand and keeps his arms in tight for a fast rotation.

keeps him from stabbing or slicing me with his knife. Though just 0.57 s elapsed since Figure 8.14, that was enough time for me to gain control over my attacker. Note that having my right leg back makes it easy for me to clamp my right arm behind Mr. Abercrombie's neck as I deliver my right knee into his solar plexus. I would follow that initial blow with three or four more combatives before fleeing the danger zone. And I would hope that my attacker doesn't have a gun, mates coming to help him, or abs of steel that would render my knee strike a failure.

I'll end this section with a quick example of how lowering your moment of inertia will help you spin faster. Look at Figure 8.16. I was initially behind Mr. Abercrombie with a rubber practice knife at his throat. He dropped quickly and rotated out of the predicament. Note his leg placement helps maintain his balance while pulling down on my arms helps disrupt my balance. Keeping his arms in tight helps keep his moment of inertia low enough that he can rotate with a lot of angular speed. At the point in the action shown in Figure 8.16, Mr. Abercrombie can stab me in the side with my knife. He could also continue his rotation and get me in side control, where he could deliver punishment.

Time to Terminate the Chapter

This chapter's concluding section is not easy to write. I'm going to discuss a very serious and extremely dangerous technique. I've already given in to the temptation to turn a phrase for the section title. When we practice with weapons like guns and knives in Krav Maga class, we tend to joke a bit more. Humor seems to lighten the gravity of what we practice. People new to a Krav Maga class usually find themselves laughing the first time someone comes at them with a practice knife or holds a training gun to their heads. Nervous laughter is inevitable when we first practice a technique that, if ever needed in real life, would mean we were in a harrowing confrontation. I laughed my first few times that I trained to defend myself against a weapon. There's nothing wrong with that. Even though training with weapons is ultimately about being prepared for the worst, releasing nervousness and tension with laughter is okay.

I did find that, as I moved through my Krav Maga training, I was less likely to laugh when training with weapons. Once the newness wore off, I found that I could focus better on learning the techniques. I'm glad that I've never been attacked by someone wielding a knife or a gun, but I hope that I could defend myself against such an attack. I can't predict with complete certainty how I would react. I do know, however, that I feel a bit more confident now that I'm past the point of nervous laughter and able to train with a more serious mind-set.

I wish to end this chapter with a termination technique. And by "termination," I do mean ending someone's life. We are taught that the techniques we employ in an actual confrontation should be proportional to the threat we face. There are situations, of course, in which we can't know the extent of the threat. If someone grabs us from behind, we don't know the intent of our attacker. Robbery? Physical harm? Sexual assault? Kidnapping? Murder? Not knowing an attacker's intent doesn't mean that the first thing we do to counter an attack is to try a termination technique. Someone whose sole purpose in messing with you is to steal your wallet doesn't deserve to die, no matter how angry you might become at the thought. If your response to an attack is totally out of proportion to the attack itself, you could be criminally liable.

So, when is termination an option? An ideal response to an attack would be to counter quickly with a couple of defensive tactics, and then evacuate the danger zone quickly. If you can't do that for some reason—your attacker is a skilled fighter, or your defensive tactics didn't have the desired effect and your attacker is still coming after you—further engagement is necessary. Termination becomes an option if you feel your own life is at risk or, at the very least, you face the threat of significant bodily harm. Maybe you've been cut by a knife and your assailant tries to cut your throat. Maybe you see your attacker pulling out a gun after trying to get you to the ground. Keep in mind that, if your attacker takes the confrontation to the level of your possible death, that person relinquishes his or her right to have his or her life protected from you.

With regard to termination, we are told, "These techniques are used to cause internal decapitation or cervical dislocation by rotating the neck at an angle beyond its normal range of movement. This action disrupts the medulla oblongata by separating the brain stem from the spinal cord. As a result the respiratory and circulatory systems of the body will cease."[5] Pretty grisly, isn't it? The first question that popped in my mind when I was initially introduced to termination techniques was, "How in the world do we *practice* this?" Maybe that question has been on your mind for a minute or two now. Krav Maga training obviously has to be done with safety in mind, but we have to be taught termination techniques with a partner simulating an attack. The key is to go *slowly* the first several times through a technique. Rotate a partner's head very slowly and release before the rotation angle gets big enough to cause even the slightest discomfort in your partner. As you get better trained, you have more control and you can move through the techniques faster. But you still need to release your partner's head before his or her head rotates very much. The idea is to get the motion down and simulate turning a head while your part-

---

5. Warrior Krav Advanced Phase A Certification Manual.

ner is very close to you, but only "follow through" with speed and aggression while your hands are off of your partner.

I had Mr. Abercrombie demonstrate one of the Krav Maga termination techniques we are taught. There are a few different ways to get an attacker on the ground after an initial assault. If one of those ways has your attacker sitting in front of you, a termination option presents itself. Only if we feel that our lives, or the lives of others, are at risk do we follow through with a termination technique. As the faux attacker, I could be reaching for a gun. Mr. Abercrombie determines that my belli-cose behavior, aggression, and vocalized threats pose lethal threats to himself and those around him. He employees a ter-mination technique as a last resort to maintain his life and the lives of others. Figure 8.17 shows me on the ground with Mr. Abercrombie behind me. He secures my chin with his right hand. His left hand prepares to move behind my head.

Figure 8.18 shows that Mr. Abercrombie has moved his left hand behind my head. His right hand has a firm grasp on my chin. My head has been rotated, essentially cocked, from the configuration in Figure 8.17. Note, too, that my head has been pushed forward. That puts my head in a "flexion" configuration (when the head moves backward, that's called an "extension" configuration). It turns out that the ligaments holding the cer-vical vertebrae together are most easily torn when the head and neck are in flexion. The space between the base of the skull and the first cervical vertebrae opens up when the head rotates for-ward. Once the connective tissues connecting the cervical verte-brae have been broken, dislodging those vertebrae and severing the spinal chord are then comparatively easy. It takes more ef-fort to rotate our heads side to side than, for example, an owl, which can nearly rotate its head 180° to look behind itself. By cocking my head, Mr. Abercrombie maximizes the angular range of motion through which he can exert a torque on my neck. That maximizes the amount of work he can perform on my neck, which maximizes the rotational kinetic energy of my neck.

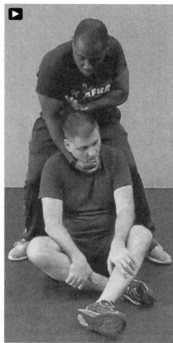

Figure 8.17
Mr. Abercrombie initiates a termination technique.

Figure 8.18
My head and neck are in flexion and my head is cocked and ready for termination.

Figure 8.19
Were this real, it would be one second before my life was extinguished.

By the time the termination sequence has evolved to Figure 8.19, Mr. Abercrombie would, in a real termination event, be rotating my head very fast. You can easily see my head in flexion. His left hand not only maintains the flexion position, it aids in rotation. The ligaments in my neck are already stretched while my head is in flexion. The added torque from Mr. Abercrombie puts my life at risk.

Now we come to the most important part of this section. If you are training and trying a termination technique meant for a real life-threatening situation, do *not* try to rotate someone's neck quickly or with much force. Safety is of paramount importance while training. Go through the motions very slowly and deliberately so that you learn the technique. Only after much practice will you be able to do what Mr. Abercrombie does in the photos shown in this section. Figure 8.20 clearly shows that Mr. Abercrombie's right hand has left my chin and has begun sliding along the right side of my head. His left hand has released the back of my head and is moving away from my head while still moving forward. The idea is to release a partner's head well before any possible harm can be inflicted on your partner.

Training is serious business, and you must continue the motions of a lethal technique so that you are comfortable with the entire motion. Figure 8.21 shows that Mr. Abercrombie has followed through with the termination technique. To analogize to something much less scary, those of you who play basketball are taught to have good follow-through on a shot. You should recognize that follow-through plays no role in a basketball's flight after the ball loses contact with the hands. What follow-through does is ensure proper technique *prior to* the ball leaving. The same applies with the termination drill. You must follow through and continue the motion to ensure that if you ever had to perform the technique in a real confrontation, you would be exerting maximum torque during the critical head rotation. Think about what would happen if you constantly prac-

Figure 8.20
Mr. Abercrombie is an experienced martial artist who knows when to release my head during training.

Figure 8.21
Mr. Abercrombie follows through on the termination technique after having released his grip on my head.

ticed the termination technique by stopping your arms immediately upon releasing your partner's head. To bring your arms to a stop, you would have to slow them down while your hands were in contact with your partner's head. You would be conditioning yourself to abate the torque you're applying to your attacker's head instead of increasing the torque. Proper training means being able to follow through with fast-moving arms, but doing so while your hands are not in contact with your partner. I know I have been a bit repetitive in this section, but that's been my intention. If you wish to practice a termination technique, please do so in the presence of a trained professional martial artist. Please do not look at the photos in this section, throw the book down, and then find a friend with a "Hey, let me show you something" offer.

Using Weapons

Some will think that using weapons is really cool. Some will be terrified to use a weapon. For those in the latter camp, there is not only dread over getting cut or shot, there is fear that pulling out a weapon, and then seeing an attacker pull out a weapon, means a life could be lost. I don't want to sugarcoat the reality of fighting with weapons. If sticks or knives are used, you will get hit or cut. There is simply no avoiding an injury, even if you are facing someone with only a modicum of skill. My guess is that those who think weapons are cool would not have the same gusto if faced with a real attacker on a dark sidewalk. Krav Maga gives you options with weapons, and with enough training and practice, those options may give you a chance you never would have had otherwise. A rare situation for sure, but if it's kill or be killed, I think all of us hope to stay alive.

As I've noted before in this book, you won't be thinking about physics when your life is on the line. But a practical understanding of physics may inform your training, and give you more confidence in what you've learned. I certainly don't claim that physics could take your training to the next level and save your life someday, but I do think of it as one of our "weapons of opportunity."[1]

What Makes a Knife Sharp?

Have you ever thought about the question that opens this section? Perhaps your answer is, "A knife is sharp when it gets sharpened." But that's completely circular. You may know that rubbing a knife on a stone or a piece of roughened metal will

---

1. Warrior Krav Level 1 Certification Manual.

sharpen the knife, but why is the knife sharp after all the rubbing? The answer lies with the idea of pressure. I discussed pressure in Chapters 2 and 3, specifically introducing the concept that pressure is a perpendicular force divided by contact area. By employing friction through the rubbing motion of a knife's edge with stone or metal, the knife's potential contact area gets to be very small. A dull knife is one that has too large a potential contact area. Look at a butter knife the next time you're eating Thanksgiving dinner and note how wide its "blade" is. A dangerous knife has a tiny potential contact area, meaning if it runs across skin with even modest downward force, the pressure can easily exceed the 100 psi required to puncture skin (see Chapter 3). The science of sharpening a knife is surprisingly complex. The material the knife is made from is important, as is the cross section of the metal. Measuring the angles made by the converging edges requires equipment more sophisticated than a protractor. Without going into all the details, I can tell you that a sharp knife has an edge thickness of a few thousandths of an inch.

Let's do a quick calculation. Take the thickness of a knife edge to be 0.004 in (four thousandths of an inch). That is roughly the thickness of a typical sheet of paper, which is why it's easy to get a paper cut. What makes a knife much more deadly than paper is that once a piece of paper's edge breaks skin, that's it. The paper is too flimsy to allow for much more force than what was necessary to puncture skin. That doesn't change the fact that a paper cut hurts, but it's a lot better than being cut with a rigid knife. Suppose 2 in of a knife's edge are in contact with a person's skin. Multiplying the width by the length gives a contact area of 0.008 in^2. If 100 psi of pressure will puncture skin, a perpendicular force of just 0.8 lb is all that's required to achieve that pressure. That force is close to the weight of three apples—it's not much. That's why a slicing knife is so dangerous. All it needs to do is make contact with skin with a tiny bit of force and the skin will be cut. If an attacker is able to get close to a victim and slice with a large amount of downward force, well in excess of 0.8 lb, the victim will not only have cut skin, but severed arteries, veins, muscles, tendons, and ligaments—that's well beyond punctured skin. A cut person also has to worry about blood loss that could result in unconsciousness or even death.

Slice and Dice Time

Krav Maga trainees are taught early on that "all attacks are targeted to vulnerable parts of the body such as the eyes, nose, ears, throat, groin, knees, and spine."[2] How do you practice slicing and dicing to the aforementioned vulnerable areas? You may use a training knife with a rubber blade, or you may opt for a training knife that provides a more realistic feel, one that has a very dull metal blade. Check out Figure 9.1. The rubber knife isn't going to hurt anyone, though I'd avoid sticking it in someone's eye. The knife with the dull metal blade isn't likely to hurt anyone either, but you must be careful.

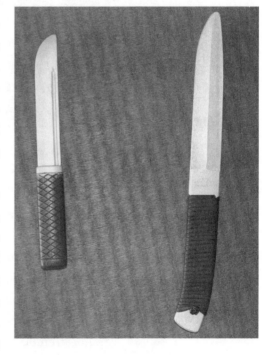

We begin with "tactical knife," which is defined as the "offensive tactical use of a knife in self-defense."[3] Note that we're talking about the *offensive* use of a knife. If you are facing an assailant with a knife, you neither know the attacker's intent nor the attacker's skill level. If you've purchased a knife for protection and find yourself in the unfortunate situation of having to use that knife for protection, you may have to counter an assailant's attack, but then follow quickly with your own counterattack. You have to go on the offensive. A person willing to attack you with a potentially deadly

Figure 9.1
Krav Maga cutlery. A rubber training knife on the left and a dulled-edge metal training knife on the right.

2. Warrior Krav Level 1 Certification Manual.
3. Warrior Krav Advanced Phase A Certification Manual.

Figure 9.2
Dr. Moorman comes at Mr. Abercrombie with a knife in an ice-pick stab approach. Mr. Abercrombie has his knife out and is about to execute Tactical Knife Entry Number One.

weapon may have to be subdued with a level of force appropriate to the threat, meaning you may have to make use of *retzef,* which is "continuous and aggressive combat motion to overwhelm the opposition."[4]

Keep in mind that the vulnerable areas I mentioned above are certainly targets, but don't expect to be so precise while using a knife. Your attacker isn't going to stand there and let you pick your favorite place to stab. That's why we go through a series of stab and slash movements. If a stab fails to do any damage, we have a slash waiting on deck. One must also be on the lookout for additional weapons. Knives are easy to conceal on one's person. Disarming an assailant might give you a rush, but if you're not careful, your attacker might have a second knife in hand, ready to do damage.

Consider an attacker coming at you with a knife, or even an ice pick. Your assailant has seen too many teenage slasher movies and comes at you with the knife up high, stabbing down. In Figure 9.2, Dr. Moorman comes at Mr. Abercrombie with a knife over her head using an ice-pick stab. She holds her knife using a forward grip such that the blade is on her thumb side. Mr. Abercrombie has pulled his knife out and is ready to counterattack, and then use his knife for offense. He holds his knife with a reverse grip, meaning his blade is on the pinkie side.

Recall that linear momentum is mass times velocity, and angular momentum is moment of inertia times angular velocity. To change the former requires a force; to change the latter

4. Warrior Krav Level 1 Certification Manual.

requires a torque. Mr. Abercrombie's first priority is getting out of the way of Dr. Moorman's knife. Look at Figure 9.3. Instead of trying to block the down-swinging knife, which would have required a decent-sized force and torque, Mr. Abercrombie stepped to his left and parried the knife to his right. All he needs to do at the start is get out of the way of the knife and put himself in position to counterattack.

In Figure 9.3, note where Mr. Abercrombie's knife is while he parries Dr. Moorman's knife. He has simultaneously parried her knife and slashed her right forearm. Were that a real knife fight, it's possible Dr. Moorman wouldn't have felt Mr. Abercrombie's knife slash her right arm. The adrenaline rush she would have had during the attack likely would have prevented her from feeling his knife cut her.

Immediately following his parry and slash, Mr. Abercrombie reverses his knife direction so that it moves in a stabbing motion. But he doesn't stab Dr. Moorman. As Figure 9.4 illus-

Figure 9.3
Mr. Abercrombie has stepped to his left, parried the knife, and slashed Dr. Moorman's right forearm.

Figure 9.4
Mr. Abercrombie has now hooked his assailant's knife.

Figure 9.5
Mr. Abercrombie initiates side control—with extra persuasion.

Figure 9.6
Side control.

trates, Mr. Abercrombie hooks Dr. Moorman's right wrist. He is out of the danger zone from her swinging knife, and now he has control of her weapon.

Mr. Abercrombie might as well make use of the linear and angular momenta in Dr. Moorman's right arm. Note in Figure 9.5 that Mr. Abercrombie has released his knife hook from Dr. Moorman's right wrist and allowed her arm's momenta to do its thing. But he didn't simply release his knife-hook grip; he moved his knife up to Dr. Moorman's throat.

Look at Figure 9.5 again and answer this question: What is Mr. Abercrombie about to do next? If you said, "side control," you are absolutely correct. Figure 9.6 shows that Mr. Abercrombie has Dr. Moorman in side control, with his knife aiding in the control of her head. But instead of administering combatives while his assailant still has a knife in her hand, Mr. Abercrombie does something really nasty. He pulls his knife from the back of her head and pulls it down the right side of her neck, as Figure 9.7 shows.

Not pretty, is it? It's not supposed to be. If someone attacks you with a knife, be prepared to get cut. Did what I just showed you seem like overkill? It would be if everything worked as smoothly as it does in training with phony knives. But in the chaos of a real confrontation, you may be lucky enough to get a single stab or slash in, and you are likely to be cut yourself. By training to follow each stab with a slash, and then each slash with a stab, you will have options. If the first move didn't work, you have an automatic reflex for the next one.

Figure 9.7
It's getting nasty now.

To summarize the knife defense and counter I just described, think about physics. The knife must be blocked or deflected in some way so that you are out of the danger zone. Mr. Abercrombie stepped aside and parried the knife. He used Dr. Moorman's momenta against her. No need to waste force and effort if you can simply sidestep and parry the knife. But note that he was always in control of Dr. Moorman's right arm—the arm holding her knife. He violently counterattacked, and used force proportional to what his attacker was about to use. During the counterattack, he also needed to be on the lookout for other weapons Dr. Moorman might pull. After he disengages from Dr. Moorman, he needs to scan the area for other attackers. There are other things to look for, as well. Are there any weapons of opportunity? Where are the safest places to exit the attack area? Once he's out of harm's way, he needs to check himself for injuries, and then seek medical attention if he's wounded. His adrenaline was rushing, just like hers, and it's possible he was stabbed and never felt it.

For all the techniques used in what I just described, physics was a constant player in the action. Besides using Dr. Moorman's momenta against her, Mr. Abercrombie maintained a wide base for balance and stability. His knife flows made effective use of his own momenta. You will get tired in an actual confrontation. There is no need to waste energy with a technique that isn't optimized by the laws of physics.

A Few Gun Basics

I discussed in Chapter 3 a self-defense technique that's used when someone has a gun pointed at your head. The one piece of physics I chose to highlight there was speed. I wish now to go over a few other gun basics that are of interest to a physicist. Then I'll look at a defense that enables a hostage to use an attacker's gun in a counterattack.

It's not hard to figure out why guns are lethal. They and the projectiles they fire are made (mostly) of metals, which are ten times denser than our skin and tissues. A high-powered rifle can fire a bullet at a speed of nearly 4,000 ft/s. Consider half that speed, 2,000 ft/s, which is about 1,364 mph, and typical of handgun bullet speeds. The speed of sound is roughly 1,125 ft/s, which is 767 mph. Without even getting specific about a particular gun's muzzle speed, you can already see one reason guns are lethal. If a gun is fired in your direction, by the time you hear "*bang*," the bullet has already reached you. The bullet is supersonic. Suppose you have a 0.5 s reaction time to seeing someone shooting at you. Ignoring air resistance, which can't be ignored in a proper trajectory analysis of a fired bullet, a bullet fired at 2,000 ft/s will go 1,000 ft in 0.5 s. That's how far from a gun you'd have to be to escape being hit, and that's only if you had the eyesight to see someone shooting at you from that distance. Even if air resistance, wind, and other factors slowed the bullet to half that speed, you'd still have to be nearly the length of two football fields away to escape being hit. And if you were able to get out of the way of the bullet, it would pass you before you heard "*bang*."

A bullet may weigh as little as a third of an ounce (mass of 10 grams). Lead has been a popular metal because it's dense, meaning it packs a lot of mass into a small space. Kinetic energy is proportional to mass, but a bullet's kinetic energy really comes from its speed. Kinetic energy is proportional to the square of the speed. A 1/3-oz bullet traveling 2,000 ft/s has a kinetic energy of 1,756 J. Double the speed and the kinetic energy quadruples. For a baseball to have 1,756 J of kinetic energy, it would have to be moving at nearly 350 mph. By the time kinetic energies reach kilojoule (kJ = 1,000 J) size, you don't want to be in the way.

Bullets are hard, and have lots of kinetic energy when they are fired. But their lethality is greatly determined by the person doing the shooting. A "good shot" is a much more dangerous person than someone who's never fired a gun. And for the last couple of centuries, the accuracy of skilled shooters has been enhanced by rifling, which is the twisting groove pattern on the inside of a gun's barrel. Those grooves help forensic scientists match bullets to guns. What they do for shooters is improve accuracy. The grooves induce spin in a bullet; when fired the bullet looks like a well-thrown spiral in football. And bullets spin for the same reason a quarterback wants a tight spiral on a football pass. A spinning bullet is much more stable than a bullet without spin. The former has significant angular momentum, which requires a net, external torque to change. Just like it's more stable to be on a moving bike than a stationary one, a spinning bullet remains stable during its flight, and thus keeps its nose pointing forward. If the orientation of the bullet could change significantly during flight, air resistance would change significantly because of the changing frontal area of the bullet. That would cause the bullet to veer off its stable trajectory.

So, these are a few tidbits of the physics I think about when considering guns in Krav Maga class. Guns and bullets are much harder than skin and tissues. Bullets travel faster than sound, which means that hearing *"bang"* won't allow you to avoid a bullet. And bullets spin when they are fired, which enhances their accuracy.

Hostage Situation

What if you are in a crowded place like a mall, and some nut who has been running from the police grabs you as a hostage? Look at Figure 9.8, in which I play the nut and Mr. Abercrombie is the hostage. I'm pointing a training handgun at someone in the crowd. I have my left arm around Mr. Abercrombie's neck. Note that he has his hands up on my left arm. That's crucial, because tenths of a second count when executing a proper gun defense. I've already told you how fast bullets fly. If the hostage can get his or her hands up without provoking a violent response, then that person needs to get his or her hands up. Even hands up in a "I surrender" gesture is better than having your hands next to your thighs. The added tenth of a second (or more) that would be needed to raise your arms during the defense could mean the difference between life and death for someone in front of you. As I describe the defense, I hope you'll note all the bits of physics we've already discussed in this book. I won't spend much time on those, because I want to get to the instant when the gun's hardness becomes an asset for the hostage.

Mr. Abercrombie's first priority is to secure the gun and get it to point in a (relatively) safe direction. With his hands upon my arm, he doesn't have far to move. But he has to move quickly. Figure 9.9 shows that Mr. Abercrombie's right hand has grabbed my right wrist, which I might otherwise swivel so as to fire in different directions. Mr. Abercrombie's left hand is moving upward with the goal of securing the underside of my gun. Note, too, that his knees are slightly bent. He is in the process of lowering his center of mass for what comes next. He'll be more stable, and he'll be in a good position to rotate. Mr. Abercrombie wasn't moving at full speed while he demonstrated the defense. But he still only took 0.4 s to get from

Figure 9.8
I've gone rogue and taken a hostage.

Figure 9.9
Mr. Abercrombie has to move fast before I fire.

Figure 9.8 to Figure 9.9. That is faster than my reaction time and, if he's lucky, I will be slow to interpret his initial hand movements as anything other than the flailing of a panicked hostage.

In Figure 9.10, you can see that Mr. Abercrombie's left hand has secured the underside of my gun. His right hand remains on my right wrist. By rotating to his left, Mr. Abercrombie has changed the firing line of my gun, and it now points to the ground. The change from Figure 9.9 to Figure 9.10 took about 0.45 s. That means that Mr. Abercrombie was able to move my gun away from my intended target in less than a second. Given my reaction time, and my possible slowness in realizing that he was fighting me, I'm not likely to have fired the gun while it was still pointed at anyone.

Mr. Abercrombie's priority now is removing the weapon from me. Even though my gun is no longer pointing where I

Figure 9.10
The gun is off my target, and Mr. Abercrombie has stored a bunch of potential energy.

Figure 9.11
I'm about to lose my gun and the use of my right forefinger.

wanted it to point, it is still dangerous. If we were standing on a hard surface, like the floor inside a mall, a shot to the ground could ricochet and hurt someone. The bullet will lose energy upon colliding with the ground, but the rebound speed can still be large enough to maintain the bullet's lethality.

What comes next is more speed and aggression. Note how Mr. Abercrombie's knees are bent in Figure 9.10. His center of mass has been lowered in the process of securing my weapon with his left hand and my weapon hand with his right hand. The process of twisting to his left and lowering his center of mass has allowed Mr. Abercrombie to store a large amount of potential energy. Much like the spring I discussed in Chapter 7, Mr. Abercrombie's tendons and muscles have been stretched and compressed in various ways so that he's like a compressed spring, ready to be released. With a slight pause in Figure 9.10, Mr. Abercrombie releases a significant amount of his stored potential energy and converts it to kinetic energy. Figure 9.11 shows that Mr. Abercrombie has rotated to his right in a move that took less than 0.7 s. By elevating his left elbow, Mr. Abercrombie is able to get the leverage he needs to rotate my gun out of my hand. In doing so, he'll likely break my trigger finger, which is why we *never* keep a finger on the trigger while training.

Beginning in Figure 9.8 and ending with the removal of my gun shortly after the scene depicted in Figure 9.11, Mr. Abercrombie was able to wound me and pilfer my weapon in 2 s. Even though he took my gun, he cannot assume that I'm no longer dangerous. I may have another weapon, such as a knife, in my pocket. To further disrupt my balance, Mr. Abercrombie makes use of my gun as a cold weapon, which means he will strike me with it. We are taught to *never* strike with the butt of a handgun. If the gun accidentally discharges, the bullet would be heading toward us. Figure 9.12 shows Mr. Abercrombie simulating a strike with the muzzle of the gun. If the gun should go off during his strike, I'm the one who will take the bullet. Having my gun taken from me and then struck hard with its muzzle will surely disrupt my balance. Mr. Abercrombie will then ensure the clip is secured in the gun and cock it to make sure a shell didn't get caught during the struggle. He will then likely hold my gun in the retention position, which means the "gun is held close to [the] body to prevent anyone from grabbing the weapon."[5]

I discuss more techniques with guns in the next chapter, only because they pertain to the advanced levels of Krav Maga. I told you already that working with weapons in class the first few times might have you nervously giggling. That's normal

Figure 9.12
I've been hoisted by my own petard.

---

5. Warrior Krav Advanced Phase A Certification Manual.

and fine, but even before you are used to training with weapons and the giggles disappear, please always treat weapons as if they are real. It might be tempting to play around with a rubber knife or a plastic gun. I admit to experiencing such temptation when I was first handed training weapons by my Krav Maga instructor. I fought the urge to play because I heeded the advice that Mr. Abercrombie passed along to the class: "Always treat training weapons as if they are real." The idea is to develop good habits. Get used to being careful while simultaneously developing confidence when you use weapons. Don't let an actual confrontation be the first setting in which you will be serious with weapons. Instead of giving in to nerves or giggles early in your weapons training, think about some of the physics ideas you've learned from this book. Focus on the ways physics explains the effectiveness of the offensive and defensive techniques you use with weapons.

I'll leave you with one last thought in this chapter. If you live in the United States, you know that guns are everywhere. Even if you don't own a gun and don't feel motivated yet to sign up for self-defense classes, at least take a short training course on guns that includes gun safety. I'm a peace-loving person, but I'm well aware that if I'm ever assaulted, a gun is likely to be present, if not used. At the very least, I want to know the basic mechanics behind firing a gun and what makes for good gun safety. I obtain gun information and learn skills with guns in my Krav Maga classes, as you will if you enroll in such classes.

Do Whatever It Takes

We have now reached this book's last chapter followed by my Final Words. I've thrown physics at you, and broken down many Krav Maga techniques. In doing so, I've set up a few scenarios that, I hope, allowed you to imagine how you might actually need to use what I discussed. I especially hope that you see the crucial role played by physics—and other sciences, too—in making Krav Maga techniques successful. The scenarios I described are certainly harrowing, whether you want to admit it out loud or not. I have no difficulty telling you that I would be terrified if someone came at me with a knife or a gun. Even some drunkard trying to fight me outside a bar would spook me. All I can hope is that I would have my wits about me, and be able to focus on what I had to do to stay safe and stay alive.

As scary as the scenarios I've described may be, I now want to up the terror level a notch or two. Even though the world as a whole has never been safer,[1] the ability to see news from around the world in an instant confronts us with some truly horrific human experiences. Gas station robberies in the news are nothing compared to the activities of terrorists. We don't help prepare people by calling terrorists "cowards," which they are clearly not. Highly motivated people[2] determined to kill innocent people aren't cowards. Calling them cowards can only cause us to lower our guards, if only very slightly, in the unfortunate event that we are confronted by terrorists. Not only are those people motivated, they are conditioned to fight. Whether

---

1. Steven Pinker's *The Better Angels of Our Nature: Why Violence Has Declined* (Penguin Books, 2012) makes a strong case for this.

2. They could have benefited the world much more as science teachers, rather than taking the road of hate, ignorance, or dogma.

it's a terrorist or a gas station robber, we should never underestimate the skill level of our attacker.

Let's now move on to some of the material covered at the black belt level in Krav Maga. What I discuss in this last chapter will include a hodgepodge of Krav Maga goodies that symbolize the essence of Krav Maga, namely a martial arts system that emphasizes the need to "Do whatever it takes." Elegant karate katas are wonderful to practice and to witness. However, real fighting does not allow you to follow one sequence of moves that were rehearsed ahead of time. Sure, developing automatic reflexes is important. But having many options at your disposal is also important. If you are used to following a jab with a cross, it will be helpful to know that if, after throwing a jab, your opponent moved too close for a good cross, an uppercut has been practiced. Forget the drunk guy stumbling out of a bar who bumps into you and wants to pick a fight. What if a terrorist has control of you?

The Bad Kind of Hoodie

A terrorist, as the term implies, wants to propagate terror. Perhaps the scariest situation for a person to face, setting aside anything harmful being done to a person's child, is being kidnapped. Not knowing who has you, where you are going, or what your abductors intend to do all contribute to the dread felt by a hostage. Check out Figure 10.1. That is me with my hands bound and my head covered with a hood. My instructor, Mr. Abercrombie, is being quite rough with me. Besides holding me with a strong grip, Mr. Abercrombie lifted me up, tossed me around, pushed me, and led me in various directions. The point of the drill is to disorient me, make me dizzy, and bring me to the point of panic. Even though I *know* that I'm in a safe training environment, and even though I *know* that my professional and talented instructor will not hurt me, the experience of being hooded, bound, and roughed up is very unsettling. It's not suited for beginning students of Krav Maga. In training for lower belts, we are acclimated to being attacked, but the severity of the attack is ratcheted up over time. If a person attending his or her first Krav Maga class were suddenly hooded, bound, and tossed about, it's possible that person might not come back for a second class.

Please take a good, long look at Figure 10.1. Do you see anything good in that image? Think physics and Krav Maga! My legs aren't bound, which is a huge plus. I have the ability to kick and run. What else? My hands are bound *in front* of me. I'm not that flexible, and if my hands were bound behind me, I could not easily get them to my front.

Figure 10.1
I am starting to wonder if Krav Maga was the best topic for my second book.

Thinking of what's depicted in Figure 10.1, do you see anything else that might be helpful to me? There is a wall behind me, which might allow me at some point to pin my abductor against. I have cropped the photo, but there is just one abductor holding me. In a real situation, there could be many assailants around me. But if I have to fight my captors, I'll take one over more than one. My hands are bound with duct tape, which wasn't heat resistant. I had a great opportunity to free my hands. Physics was on my side. I knew that if I created enough friction between my hands and the tape, I could heat up the sticky adhesive on the tape. While I was awaiting the right moment to fight back, I kept moving my hands back and forth, which created friction with the sticky part of the tape, which created heat flow into the glue, which lost most of its stickiness, and that allowed me to pop off the tape. What if Mr. Abercrombie had used better tape? What if I couldn't easily free my hands? Think using physics. I have my hands in front of me, and they're balled into fists. I can fight with a single, big fist, and I can swing my arms rather easily because of relatively small moment of inertia. I even have the opportunity to get my arms around my attacker's head.

When faced with a life-or-death situation, do everything possible to avoid thinking, "Well, gee, I'm in deep trouble and about to become room temperature." As difficult as it is to imagine, try to focus on your attacker and on your surroundings. Think about what you can use to your advantage. If you can't see anything, listen closely to any sounds you pick up. Are you in a place with an echo, like a warehouse? Can you hear traffic? Do you hear anything familiar from the weather, like rain or wind? Can you hear other voices and, if so, are they aggressive like abductors or scared like hostages? What does the ground feel like? When I was being attacked by Mr. Abercrombie, I obviously knew I was training on padded floor mats, so I may not have been worried about hitting the ground. In a real situation, see if you can tell if you are on concrete, grass, dirt, carpet, and so on. What do you smell? Any type of fuel? Smoke or gun powder? Use your senses; we can assess our environment quite well without our sight.

Now I need to let you in on something that made what I went through in Figure 10.1 especially stressful. The hood I was wearing was doused in water before it was placed on my head. Though that didn't really put me in a position of being waterboarded, it did move me in that direction. We humans have come up with some truly disgusting ways to treat our fellow humans. In actual waterboarding, a person is placed on a table that's inclined such that his or her head is slightly below his or her feet. The victim is usually hooded, which makes knowing what's about to happen a challenge. Water is poured onto the hood over the person's nose and mouth. The person almost immediately gags. I remember reading about the experience of being waterboarded—and watching that experience—in a piece written by the late, great Christopher Hitchens.[3] It's a terrifying procedure to read about, and it was nauseating to watch. It really is drowning, and it's definitely torture.

Water is a fascinating molecule. It dominates who we are, and our existence demands that we have access to it. Between 50% and 60% of our body weight is water. For our purposes here, I'll tell you about water's "surface tension." You've surely noticed drops of water on a shower door, and on the blades of grass that make up morning dew in your yard. Ever wonder why water exists in drops, sometimes nearly spherical in shape? Water is, after all, a liquid, and it should flow with the help of gravity, right? Why does it stick to objects, and even hold a nearly spherical shape? The answers to those questions lie with surface tension. Without going into all the fascinating details underlying the phenomenon of surface tension, just think of it as water having the ability to maintain a sticky membrane that can be under tension. Certain small bugs can walk on water because of surface tension. People used to make compasses by carefully placing a tiny piece of metal on water and letting Earth's magnetic field do its thing with the metal. Metal is almost ten times denser than water, yet surface tension allows a thin piece of it to float on the surface of water. Surface tension can, of course, be broken if the force applied to the water is strong enough. That's a good thing, or we couldn't jump into swimming pools.

Water drops form spherical shapes because, for a given volume of water, a sphere minimizes surface area, which minimizes energy. Those spherical drops get stuck on top of some of the nooks and crannies of clothing and other material. When clothing gets wet, the water doesn't just fall off the fabric. Surface tension helps water adhere to the clothes. If the article of cloth-

---

3. See "Believe Me, It's Torture" in the August 2008 issue of *Vanity Fair* and accompanying video.

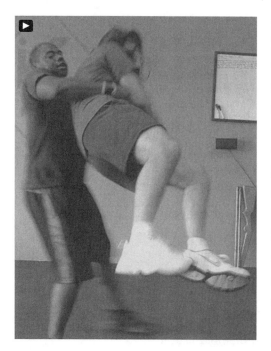

Figure 10.2
I'm getting tossed about by my captor.

ing happens to be a terrorist's hood, water has the unfortunate property of clogging up openings in the fabric. That obviously makes breathing difficult, because air can't easily pass in and out of the fabric's tiny openings. I most definitely noticed this while donning the soaked hood. Unlike having a dry hood on my head, which I tried, I had enormous difficulty pulling air into the wet hood when I attempted to breathe. Raising the temperature of the water is one way to decrease surface tension, but that's not necessarily a great option if you are fighting for your life. In ordinary life, compounds called surfactants, which are contained in laundry detergent, help reduce surface tension. That's what allows water to lose some of its stickiness and make its way into all the nooks and crannies in clothing, thus flushing out dirt and grime. But it's not likely that a terrorist will put soap or detergent on a hood before dousing it with water.

I've tried to identify some positives in a seemingly hopeless situation. I've identified some obvious problems, and other not-so-obvious problems, like a soaked hood. If you're ever in a situation like what's depicted in Figure 10.1, you are sure to be terrified, off-balance, struggling to breathe, and uncertain of what's coming next. Your only chance of survival could hinge on summoning the fortitude to take stock of your surroundings and pounce on any opportunity that presents itself. Do you see any opportunities in Figure 10.2? I look to be in a world of hurt, don't I? I did try my best to keep the room's orientation in my head while I was being tossed around. Put your physics cap on and think about opportunities, no matter how slim, that I might exploit. We all agree that my balance has been thrown out of whack. But look at Mr. Abercrombie. He has to lean back to maintain his balance. If I may pilfer a phrase I adored when I read Chapter 58 of Herman Melville's *Moby-Dick*, hauling my "bulky masses of overgrowth" into the air means Mr. Abercrombie has to fight a torque caused by the gravitational force on me. Sending my head back doesn't look like an option, so I may have to get to the ground before I have a solid opening. But despite Mr. Abercrombie's efforts to terrorize me, he is outputting energy, which will tire him, and he's disrupting his balance.

I actually tried something on my own during my hooded training exercise. I was thinking physics, and wondering if

Figure 10.3
Note what I did with my legs.

there was something I could do to make it more difficult for Mr. Abercrombie to toss me around. And then the term "moment of inertia" hit me. There wasn't much I could do with my arms because my hands were still bound, and I felt like I needed to protect my torso. But remember that my legs weren't bound. So look what I did in Figure 10.3. I spread my legs and pushed them away from Mr. Abercrombie's body, thereby increasing my body's moment of inertia. I could immediately notice Mr. Abercrombie struggle a bit to keep me spinning. Having my legs apart had the added benefit that when Mr. Abercrombie finally put me down, I had a wide enough base to have good balance. If all it did was tire my abductor a little more, it might make all the difference when the time came to fight back.

The training exercise then moved to fighting. My hood was removed, Mr. Abercrombie grabbed a body shield, and I was supposed to unload on him. I took advantage of my arms' low moment of inertia and hit the body shield with both fists. Figure 10.4 shows me with my hood mercifully off, fighting my captor. Note the wide base I have to maintain my stability. Mr. Ab-

Figure 10.4
It's time for me to fight back.

Figure 10.5
Fighting isn't limited to my arms.

ercrombie isn't just letting me hit the body shield; he's giving me hard feedback and pushing me when I don't punch. And never think that fighting is limited to arms and hands. In Figure 10.5, I managed to get my hands behind Mr. Abercrombie's right shoulder and deliver a knee strike. I would normally use my right leg for such a strike, but in the terrorist drill, I used both legs. Remember *retzef* from the previous chapter? If it's kill or be killed, continuous and aggressive combat is what's necessary.

I was by no means finished with the terrorist drill following a few hits on the body shield. I was hooded again, tossed about again, had the hood removed, and had to fight again. I was pretty drained when it was all over. But all the while, I kept sound physics applications in my head, like wide base, moment of inertia, and balance disruption. Before I was hooded again, I managed to break free of the duct tape that bound my hands. During my second round of fighting, I was able to punch and get Mr. Abercrombie into side control.

Don't think that what I've described in this section will guarantee you a safe exit from a terrorist attack. Even in the upper belts, we exchange a few nervous jokes while doing terrorist drills. Just like any attack, a terrorist attack won't be scripted. Unlike a shove outside a bar, however, a terrorist attack puts lives of innocent people at risk. We are taught in Krav Maga class to keep our wits about us, and if we see an opportunity to turn the tables on an abductor and fight back, we take that opportunity. We take it like our lives are at stake, because they likely will be. If the chances of surviving an abduction are 0% without training and 2% with training, which option do you want?

If an opportunity to fight back doesn't present itself, or if you're too terrified to even attempt fighting back, at least pay good attention to all the details of your surroundings. Those details could save your life or the lives of others. Do you see any exits? Have you counted how many terrorists there are? Can you ascertain what types of weapons they are using? Are they communicating on cell phones or other devices, which suggests that they are getting help or working in conjunction with the pernicious activities of their friends? Do you see vehicles? If so, can you make out license plates, tire treads, makes and models, colors, and so on? Do you recognize the language spoken by your captors? There are many other questions I could pose here, but I don't want to digress from Krav Maga science. Sign up for a Krav Maga class and you'll learn a lot more about winning the mental game in a hostage or terrorist situation.

A Cold Long Rifle

In the United States, it is comically and obscenely easy to acquire a weapon designed for mass killing. We may not see the world's most used assault rifle, the AK-47 (Kalashnikov), in local news too often, but weapons like the AR-15 are seen regularly in the news in conjunction with mass shootings. Those in the military may scoff at the idea of using an AR-15 in combat, but it doesn't change the fact that this weapon is capable of taking many lives quickly, and there are at least 10 million of these guns in the

United States.[4] This book is not the setting for a gun debate, or to ask why we in the United States haven't had our Dunblane or Port Arthur moment yet.[5] I want to stick to how physics can help in Krav Maga class, and one of the topics we deal with is assault with a rifle. There are many kinds of rifles. In our classes, we focus on drills called "Tactical Long Gun," which is the "offensive use of a long gun in self-defense."[6] What we cover applies to most any type of rifle. The idea isn't to get too pedantic with the name, but to separate the offenses and defenses we practice with a rifle compared to those with a handgun.

So here's a scenario that could lead you to engaging in self-defense with a long gun. You and some friends are in a crowded club. The music is loud, overhead lights are dim, and colorful moonflower lights dance across the club in step with loud music and patrons' dancing. It's noisy; people are having a good time getting a buzz on and socializing. You suddenly hear the distinct sound of gun fire. You hear people screaming. In seconds the club is sheer chaos. People are running everywhere, and you finally catch a glimpse of a guy with an AR-15 firing in random directions. In the blink of an eye, your fun evening out now has you square in the middle of a terrorist attack.

I can't imagine anyone wanting to be in the aforementioned scenario. There will clearly be people running away. Survival instincts kick in, and people want to flee danger. Some people, whether they've never been in a fight or they're trained military personnel who've seen combat, will be in the wrong place at the wrong time and be killed. Maybe you've had some Krav Maga training, and upon hearing gunshots and screams, you ducked for cover, kept your wits about you, and took in as much information as possible. You feel confident that there is just one gunman in the club, though that confidence can't extend to

4. "AR-15 Rifles Are Beloved, Reviled and a Common Element in Mass Shootings," Alan Feuer, *New York Times*, 13 June 2016.

5. If the names "Dunblane" and "Port Arthur" don't mean anything to you, I recommend starting with Wikipedia.

6. Warrior Advanced Krav Maga Phase B Certification Manual.

Figure 10.6
Mr. Abercrombie prepares to inflict damage on the body shield using a training long gun.

outside the club. The gunman wanders in your direction, runs out of ammunition, and prepares to reload with a new magazine. You seize the opportunity to go for the gunman's weapon. Your aggressiveness shocks the terrorist, and you manage to get his AR-15. Before he can register what's happening, you want to inflict damage so that he's subdued and no longer a threat to anyone. You don't feel that you can kill, and it's possible that you can't fire the weapon because the new magazine wasn't properly loaded. You choose instead to use the long gun as a cold weapon, which means you will inflict "blunt force trauma"[7] on the assailant.

Look at Figure 10.6. Mr. Abercrombie has the assault rifle and I'm holding a body shield, preparing for him to demonstrate cold strikes with the weapon. We train with wooden weapons, but we are taught to always assume those weapons are loaded and real. Being nonchalant or playful with the training weapons is frowned upon, and for good reason. When handling an actual firearm, we must always respect what we are

7. Warrior Advanced Krav Maga Phase B Certification Manual.

Figure 10.7

Tactical long gun with a cold weapon: (1) diagonal butt strike up, (2) diagonal barrel strike down, (3) thrust with muzzle, (4) step horizontal butt strike, (5) back butt strike, and (6) step back to shooting position. Arrows show direction that the gun's barrel is moving.

holding, and that includes treating the weapon at all times as if it's loaded. We develop good habits in class so that we're not careless with real weapons. Mr. Abercrombie began his demonstration with the muzzle of the training gun pointing down. His eyes are locked on me and he's in a wide stance that you now know represents good balance. I have one leg back in anticipation of hard strikes to the body shield. Mr. Abercrombie's goal is to move quickly through a sequence of strikes. In a real situation, that flow will hopefully be second nature, and if one strike is unsuccessful, the next one will be on the way without much thought. The sequence is this: (1) diagonal butt strike up,

(2) diagonal barrel strike down, (3) thrust with muzzle, (4) step horizontal butt strike, (5) back butt strike, and (6) step back to shooting position.[8]

I show the aforementioned six moves all in Figure 10.7. Mr. Abercrombie executed the six moves in 2.25 s, beginning with his position in Figure 10.6. If a terrorist is startled enough when his weapon is taken from him, and if you're able to land two or three strikes with his weapon, you're doing wonderfully. Now let's think about the physics involved in those six moves.

8. Warrior Advanced Krav Maga Phase B Certification Manual.

Remember that a "wide stance ensures stability." I can't emphasize enough how important balance is when confronting an attacker. Your life could be lost in a harrowing situation like a terrorist attack if you decide to confront your assailant, and then you lose your balance. Look where Mr. Abercrombie holds the gun. His hands are in the middle of the gun. That minimizes the gun's moment of inertia because the majority of the gun's mass is close to the rotation axis through the (roughly) center of the gun. That small moment of inertia means Mr. Abercrombie doesn't have to exert much torque to rotate the gun. His arms are in tight, which means his torso doesn't have to exert much torque to rotate his arms. Note in each of the five strikes, Mr. Abercrombie's eyes are dead on his target. Notice how one move naturally follows from the previous move, which means the flow takes advantage of your own momentum, both linear and angular. Once he strikes upward with the diagonal butt strike (1), he's stored potential energy in his muscles and tendons as he momentarily stops the gun over his left shoulder. When he executes the diagonal barrel strike down (2), he converts a lot of that stored energy into kinetic energy.

As with everything I've discussed in this book, I could bring in all the laws of physics to help describe what's going on with the tactical long gun sequence. But notice how many areas of physics I've already touched on in the previous paragraph. Let's keep going! The arrows in Figure 10.7 show the direction the barrel of the gun is moving. Take another look at Figure 4.4. Now look at Figure 10.7 again. Diagonal strikes (1) and (2) are both made with the end of the barrel, which is the high kinetic energy zone. Recall that a net, external force is required to change an object's linear momentum. Want to stop the two strikes (3) and (5) in Figure 10.7? You'll need to apply a force. That's what I'm doing with the body shield and my body. But note that Mr. Abercrombie steps into his horizontal strikes. That adds linear momentum to the system comprised of the gun and Mr. Abercrombie. In other words, there is more than just the gun and its mass moving toward me. Stopping the gun also means stopping Mr. Abercrombie's forward movement.

You can see how the balance of a terrorist could be disrupted by his mental mistake of not noticing the movement of the person thrusting the gun at him. I've never jousted, but I've heard from people who've tried it that the force from the lance is unexpected. Unless you have experience being hit by a lance, you can't anticipate how it will feel to be hit by one that's being pushed by a person on a fast-moving horse. Similarly, a simple step into the thrust with a cold long gun can provide enough added linear momentum to stun your assailant.

Now take a good look at the shooting position in (6) in Figure 10.7. Mr. Abercrombie holds the gun close to his body. He's trying to keep the gun from being taken from him. His stance is wide to help his balance, but note that it's his right leg that's back behind him. Because he has the butt of the rifle nestled against his right shoulder, if he does have the option to fire, the gun's recoil will be against his right shoulder. His right leg thus helps anchor his body if he had to experience significant recoil. Mr. Abercrombie has fired plenty of guns, so he's familiar with the recoil of each type. If you're unfamiliar with a gun's recoil, you should definitely position yourself as shown in (6) in Figure 10.7. You don't want your balance to be disrupted in the unfortunate event that you have to fire the weapon. Also notice that Mr. Abercrombie's eyes aren't too low on the gun. He has the option of using the sights, but he's not blocking his short-range vision. Can you think of another advantage to his stance? By turning slightly sideways, Mr. Abercrombie has reduced the cross-sectional area seen by the terrorist. If the terrorist has a handgun, his new target will have a smaller area than if Mr. Abercrombie were standing up and facing the terrorist. Mr. Abercrombie's not safe from gunfire, but his arms are partially shielding his internal organs and his head is partially obstructed by the gun. *Any* reduction in a terrorist's target area is absolutely encouraged.

Hot Weapon

What if you are in the midst of a terrorist attack and you have a weapon? Perhaps a weapon you seized from one of the terrorists, as described in the previous section? You may find yourself in a situation where you have to shoot that weapon, which means you will use the gun as a hot weapon. We practice several shooting drills in Krav Maga class. One such drill involves the shooting position sequence of standing, kneeling, and prone.[9] Before we ever fire a gun, we need to know how to hold it and how to position ourselves for firing. We always think about our balance. Transitioning from one shooting position to another is also important to practice. No real shooting situation will have you standing out in the open in your favorite shooting stance for several minutes. Once you fire a gun during a terrorist attack, you alert the terrorist or terrorists to danger, and you become more than a random target. No Krav Maga class can teach you exactly when to shoot and when not to shoot. The idea is to train you so that, when you make up your mind, you can engage the threat with lethal gun force.

I admit that when I first went through the shooting sequence, I was a bit bored. Stand, kneel, lie down. I certainly didn't feel as excited as when I shot the gun. But that's like a beginning basketball player feeling bored while doing footwork drills instead of shooting. But the coach knows what he or she is doing. And so do our Krav Maga instructors. There is no point firing the gun if we don't know how to hold it properly, and how to orient ourselves in various shooting positions. Physics is like that, too. We have to put our students through years of classwork and lab work before they are ready to go out into the world as scientists.

Figure 10.8 shows Mr. Abercrombie in the three aforementioned shooting positions. He's using the same training long

---

9. Warrior Advanced Krav Maga Phase B Certification Manual.

(1) Standing

(2) Kneeling

(3) Prone

gun that was used in the previous section. For the sequence of photos shown in the figure, he opted to show a standing form different from (6) in Figure 10.7. For those who feel more comfortable with your feet shoulder width apart instead of having one leg back, the proper stance is shown in Figure 10.8. Because the gun is "long" and his right hand is on the trigger and pistol grip, Mr. Abercrombie has extended his left arm so that his left hand is near the front sight assembly. That helps with balance, because an external torque on the gun will have a harder time rotating the gun if his hands are apart. If his left hand was closer to his right hand, the gun could be rotated more easily

Figure 10.8
Mr. Abercrombie demonstrates the shooting position sequence.

(perhaps by an assailant). Note the important change in Mr. Abercrombie's posture when he fires with his feet below his shoulders. To compensate for not having his right leg back to assist in absorbing recoil after the gun is fired, Mr. Abercrombie has pushed his derriére out. With his arms and gun extended in one direction and his backside in the opposite direction, his center of mass is over a spot between his shoes. If there is much recoil after firing, his center of mass likely won't move past his shoes. Had he been standing straight up, a recoil could push his torso back and his center of mass could move over his shoes, thus disrupting his balance, and possibly leading to a fall.

Mr. Abercrombie's right leg is back when he is in the kneeling position. Had he chosen to squat down instead of kneel down, he would have been very susceptible to rotating backward from the gun's recoil. If you have to be in a kneeling position for an extended time, resting your elbows on your knees helps prevent cramping, and it helps with stability while shooting. Some shooting situations require you to move from the standing to the kneeling position quickly, so there may not be time to get too comfortable in the kneeling position.

One important thing to note about the prone position is the orientation of Mr. Abercrombie's shoes. His frontal cross-sectional area will increase if he raises his shoes and rests on the front points. Remember to make as small a target as possible. Having your legs spread increases stability. Some shooters prefer having one leg out to the side with a bent knee. That's a fine position, too. Take special care, when getting into the prone shooting position, to put your non-shooting hand on the ground first. Steady yourself with that hand as you position your legs. Never stabilize yourself by putting your gun-holding hand on the ground first. You risk damaging the rifle. Even getting some dirt or other debris in the muzzle while improperly getting into the prone position can affect shooting accuracy.

Final Words

Our exploration of the science of Krav Maga comes to a close with this section. Have I shown you every technique in Krav Maga? Not by a long shot. Have I dissected each technique that I did show with all possible physics descriptions? No way. But you don't need me to spoon-feed you every possible example and write 100 pages on each. Physics is a joyous field in which to make a living. I hope some of that joy has rubbed off on you. If all you get from this book are a few physics concepts—say, stability, torque, and moment of inertia—I'll be thrilled. The reason you don't need me to present every Krav Maga technique is that, after reading this book, you're armed with the basics needed to describe them all. In fact, no Krav Maga instructor can present you with all possible dangerous scenarios that you could encounter. You are taught the basics, and if you're ever unlucky enough to find yourself in a harrowing situation, those basics might save your life. You'll know how to adapt to something new.

And I hope that this excursion into physics will help you to see the larger world with new eyes. When you're watching a figure skater's rotation rate skyrocket as she pulls her extended arms and leg into a spin, I'll be ecstatic if you think, "Wait, that has something to do with moment of inertia and angular momentum conservation." I tell my students that once they learn a topic in physics, they'll start seeing that topic everywhere they look. Suddenly the world becomes more understandable, and that's part of the excitement of being a scientist.

If you haven't already signed up for a Krav Maga class, do so. Or sign up for another martial arts class or a self-defense seminar. Even if you don't want to be the next Bruce Lee, you'll have fun meeting new people, you'll learn a few things that could help you defend yourself against an attacker, and you'll benefit from the exercise. And you'll see things you've read in this book in every technique you're exposed to. Instructor explanations will sink in a little better with you because you'll have a deeper understanding of *why* the techniques are so effective. If you happen to live near Lynchburg or Forest, Virginia, please stop by Mr. Abercrombie's *Warrior Success Academy*. You won't find a better instructor anywhere. You may also see a physicist trying to get his not-as-flexible-as-it-should-be, middle-aged body free from one of Mr. Abercrombie's strong holds. That will be me. Please introduce yourself. We'll chat some physics during class, and perhaps afterward over a much needed pint. Good luck with your training and *never stop learning!*

Index